AF685766

Archaeology, History and Science

PUBLICATIONS OF THE
INSTITUTE OF ARCHAEOLOGY, UNIVERSITY COLLEGE LONDON
Series Editor: Ruth Whitehouse
Director of the Institute: Stephen Shennan
Founding Series Editor: Peter J. Ucko

The Institute of Archaeology of University College London is one of the oldest, largest, and most prestigious archaeology research facilities in the world. Its extensive publications programme includes the best theory, research, pedagogy, and reference materials in archaeology and cognate disciplines, through publishing exemplary work of scholars worldwide. Through its publications, the Institute brings together key areas of theoretical and substantive knowledge, improves archaeological practice, and brings archaeological findings to the general public, researchers, and practitioners. It also publishes staff research projects, site and survey reports, and conference proceedings. The publications programme, formerly developed in-house or in conjunction with UCL Press, is now produced in partnership with Left Coast Press, Inc. The Institute can be accessed online at www.ucl.ac.uk/archaeology.

ENCOUNTERS WITH ANCIENT EGYPT Subseries, Peter J. Ucko (ed.)
Jean-Marcel Humbert and Clifford Price (eds.), Imhotep Today
David Jeffreys (ed.), Views of Ancient Egypt since Napoleon Bonaparte
Sally MacDonald and Michael Rice (eds.), Consuming Ancient Egypt
Roger Matthews and Cornelia Roemer (eds.), Ancient Perspectives on Egypt
David O'Connor and Andrew Reid (eds.), Ancient Egypt in Africa
John Tait (ed.), 'Never had the like occurred'
David O'Connor and Stephen Quirke (eds.), Mysterious Lands
Peter J. Ucko and Timothy Champion (eds.), The Wisdom of Egypt

CRITICAL PERSPECTIVES ON CULTURAL HERITAGE Subseries, Beverley Butler (ed.)
Beverley Butler, Return to Alexandria
Ferdinand de Jong and Michael Rowlands (eds.), Reclaiming Heritage
Dean Sully (ed.), Decolonising Conservation

OTHER TITLES
Andrew Gardner (ed.), Agency Uncovered
Okasha El-Daly, Egyptology, The Missing Millennium
Ruth Mace, Clare J. Holden, and Stephen Shennan (eds.), Evolution of Cultural Diversity
Arkadiusz Marciniak, Placing Animals in the Neolithic
Robert Layton, Stephen Shennan, and Peter Stone (eds.), A Future for Archaeology
Joost Fontein, The Silence of Great Zimbabwe
Gabriele Puschnigg, Ceramics of the Merv Oasis
James Graham-Campbell and Gareth Williams (eds.), Silver Economy in the Viking Age
Barbara Bender, Sue Hamilton, and Chris Tilley, Stone World
Andrew Gardner, An Archaeology of Identity
Sue Hamilton, Ruth Whitehouse, and Katherine I. Wright (eds.), Archaeology and Women
Gustavo Politis, Nukak
Sue Colledge and James Conolly (eds.), The Origins and Spread of Domestic Plants in Southwest Asia and Europe
Timothy Clack and Marcus Brittain (eds.), Archaeology and the Media
Janet Picton, Stephen Quirke, and Paul C. Roberts (eds.), Living Images
Tony Waldron, Paleoepidemiology
Eleni Asouti and Dorian Q. Fuller, Trees and Woodlands of South India
Russell McDougall and Iain Davidson (eds.), The Roth Family, Anthropology, and Colonial Administration
Elizabeth Pye (ed.), The Power of Touch
Miriam Davis, Dame Kathleen Kenyon
Marcos Martinón-Torres and Thilo Rehren (eds.), Archaeology, History and Science
John Tait, Why the Egyptians Wrote Books

Archaeology, History and Science

Integrating Approaches to Ancient Materials

Edited by **Marcos Martinón-Torres** and **Thilo Rehren**

Left Coast Press Inc.

Walnut Creek, CA

Left Coast Press, Inc.
1630 North Main Street, #400
Walnut Creek, California 94596
http://www.LCoastPress.com

Copyright © 2008 by Left Coast Press, Inc.

All rights reserved. No part of this publication may be reproduced, stored in a retrieval system, or transmitted in any form or by any means, electronic, mechanical, photocopying, recording, or otherwise, without the prior permission of the publisher.

Hardback ISBN 978-1-59874-340-1

Library of Congress Cataloging-in-Publication Data

Archaeology, history and science : integrating approaches to ancient materials/edited by Marcos Martinón-Torres and Thilo Rehren.
p. cm. — (University College London Institute of Archaeology)
ISBN 978-1-59874-340-1 (hardback : alk. paper)
1. Archaeology—Methodology. 2. Antiquities—Collection and preservation. 3. Antiquities—Analysis. 4. Archaeology and history. 5. Archaeometry. 6. Archaeological chemistry. I. Martinón-Torres, Marcos. II. Rehren, Thilo.
CC75.7.A725 2008
930.1028—dc22 2007044315

08 09 10 5 4 3 2 1

Printed in the United States of America

∞™ The paper used in this publication meets the minimum requirements of American National Standard for Information Sciences—Permanence of Paper for Printed Library Materials, ANSI/NISO Z39.48—1992.

Cover design by Andrew Brozyna

Contents

Illustrations

Figures

Tables

Preface

This book has been in gestation for a long time. Most of the contributions stem from papers presented at a conference organised at the UCL Institute of Archaeology in 2001, which were more recently developed into the essays presented here. These include the chapters by Graves-Brown, Shortland, Freestone, Ashton, Bayley, Rehren and Martinón-Torres, and Badcock and Crossley. Publication had to wait owing to a variety of practical matters, but the delay allowed us to include further relevant contributions by Pérez-Arantegui and Pardos, and by Jackson and Smedley.

The combination and confrontation of historical and archaeological sources of information is not new, nor is the application of scientific techniques to the study of archaeological materials. In fact, in the years elapsed since the original conference, the amount and quality of interdisciplinary work have continued to grow strongly. However, only rarely does one find an explicit and focused critique of the various sources and approaches employed, as well as their integration, especially when archaeometric analyses are included in the equation.

This collection of papers is particularly miscellaneous because of the wide range of cultures, periods, and materials addressed, and the diversity of starting premises, analytical scales, and methodological approaches of the different contributors. Also, the balance among archaeology, history, and science is unequal across the board. We simply encouraged authors to freely present the results of their work, but also reflections from their own experience in combining sources and research strategies. There is, however, a strong common theme to all of them. In one way or another, they all reflect on the particular merits of combining often seemingly noncompatible sources, resulting in a significantly enhanced understanding of the subject matter at hand.

The scope of the book has certain biases, though. First, most of the contributors are scholars working in archaeology departments. Second, a good number of the papers fall within the broad field of ancient technology studies – namely, the production and use of stone artefacts, ceramics, metals, or glass. These constraints are partly a reflection of the professional acquaintances and research interests of the editors, but they may be justifiable, since this is an area where archaeological, textual, and scientific analyses are routinely combined, all too often with little reflection on the variable nature and value of the different resources.

The resulting volume is thus diverse, but, we hope, rich, too, and of relevance outside archaeology and technology studies. Besides offering individual contributions to specific areas of research, the volume as a whole should prompt some general reflections on the problems and potentials of the interdisciplinary study of the human past. In this vein, the introductory chapter draws primarily on the case studies from this book, placed within a broader context, in an attempt to present a critical view and some suggestions for future work. However, rather than attempting to impose a single vision, we hope that this volume will provide encouragement for more alert and critical interdisciplinary work.

A number of individuals have been instrumental in the materialisation of this book. First of all, we should thank the individual authors for embracing the overall theme and contributing their insights but also for patiently bearing with us in spite of the delays. The expertise generously offered by a wide range of referees from different disciplines has significantly helped raise the standards of this volume. We would also like to acknowledge the editorial assistance provided by Mitch Allen, Jennifer Collier, and Stacey Sawyer at Left Coast Press and by Sue Poll and Marion Cutting in London. Finally, we remain indebted to the late Peter Ucko, who was Director of the UCL Institute of Archaeology while this book was being produced, and who, we are sad to say, passed away shortly before it was completed. He supported, enabled, and inspired us, and his memory will continue to do so.

Marcos Martinón-Torres and Thilo Rehren
London, 1 July 2007

CHAPTER 1

Why Should Archaeologists Take History and Science Seriously?

MARCOS MARTINÓN-TORRES

ABSTRACT *A significant number of studies of the human past involve a combination of historical, archaeological, and archaeometric information. However, the complex relationships between the different sources of information and research strategies are rarely considered critically and explicitly. This chapter presents a brief review of theoretical positions regarding those relationships, and of relevant methodological approaches. Subsequently, the discussion concentrates on the study of ancient materials and technologies, as a prominent field where archaeological, textual, and scientific analyses are routinely combined, often with little reflection on the variable nature and value of different resources. Based on a critical assessment of relevant case studies and broader literature, some recommendations are made for future work: These include a clearer focus on the contextual encounters between materials and texts, the definition of wider questions and theoretical paradigms engaging different specialists, and a more conscious and socially aware approach to archaeometric research and dissemination.*

Introduction

Until recently (and still currently in many fora), it seemed entirely appropriate to ask questions such as: To what extent is archaeology necessary for the study of aspects of those periods from which written sources are available? I cannot help but wonder what these enquirers would make of the opposite question: Is historical research necessary when archaeological information is at hand? Twenty years ago, Lloyd (1986) persuasively argued why historians should take archaeology seriously and ended his essay by challenging historians to address an equally valid question: Why should archaeologists take history seriously? As far as I am aware, nobody has felt the need to answer, probably because the role and the value of history are taken for granted.

When archaeometric approaches are included in this debate, the matter becomes yet more complex, not least because analytical techniques often require extra time and resources, as well as invasive removal of samples from archaeological materials. For those who question the pertinence of invasive analysis of archaeological remains, the application of scientific techniques in the study of text-informed periods might appear even less necessary.

We ought not to be pessimistic, though; the current situation is rather encouraging. There is increasing dialogue between archaeometrists and archaeologists, and between the latter and historians, as well as growing numbers of individuals who cross the boundaries between two or all three specialities. What we do miss is a more overt critical consideration of how texts, materials, and analytical data relate to one another and explicit attempts at proposing best-practice strategies for these kinds of interdisciplinary research, especially for those involving archaeometric analyses.

This chapter aims at contributing to this optimistic scenario, by raising issues that should be of interest to archaeologists, historians, and scientists. Are there any common problems for these specialists? Can we define *in abstracto* the nature and informative potential of materials and documents? Is it possible to establish an overarching paradigm that may guide future practice? By reviewing the case studies presented in this book and other relevant publications, and considering the broader context in which they are inscribed, this introductory chapter tries to envisage some way forward.

Following a brief review of theoretical positions regarding the relationships between texts and archaeological materials, and of different methodological approaches, this essay concentrates on studies of ancient materials and technologies. The justification for this is twofold: First, I believe that technological activities and materials are cultural expressions, and thus their study constitutes one of the most powerful lines of enquiry about the human past; second, given the relative abundance of technical texts and recipes in the historical record, this is a prominent field where archaeological, textual, and scientific analyses are routinely combined, though often with little reflection on the variable nature and value of different resources. It is hoped that an explicitly critical approach may prove stimulating for future work.

Texts and Materials: A Changing Relationship

In two inspiring books, Andrén (1998) and Moreland (2001) have explored the evolving perceptions of the role and the value of documents and artefacts in our reconstructions of the past. Starting with early essays blatantly claiming the banality of historical archaeologies, both reviews move on to somewhat more encouraging views that have advocated the use of both types of sources, as they allegedly inform distinct questions. Texts and archaeology have been seen as informative of, respectively, ideology and technology, elites and peoples without history, the emic and the etic, the conscious and the unconscious, and even big questions and small questions. Both Andrén and Moreland contend that these perspectives implicitly or explicitly advocate the subordination of archaeology to the empire of the Word: If archaeology is used only to fill the gaps left by more or less spaced out written sources, or as a 'testing ground' to prove or refine wider theories formulated by historians, then its potential is seriously constrained. Materials appear as illustrations of historical stories, examples, or exceptions for broad reconstructions – 'the icing on the cake' – and the archaeologist's ability to reconstruct past societies is seriously hindered. This excessive reliance on textual information or history-driven paradigms has raised colourful expressions such as 'the tyranny of the historical record' (Champion 1990) and 'text-misled archaeology' (Young 1992) and has inspired alternative programmes such as that of 'text-aided archaeology' (Little 1992).

These generic positions, however, are bound to demand qualifiers and exceptions and to raise disagreement among specialists. The way forward, according to Moreland (2001:77–97), entails beginning to appreciate the context-specific origins and meanings of both texts and materials, and their active role in the reproduction and transformation of power and identity. Andrén (1998:153–77) also advocates a more thorough 'construction of the context', an interpretation of the actual encounter between material culture and writing, and their relationship in each case. These useful perspectives will be revisited later but, before that, it is worth bringing up some of the more traditional viewpoints.

One of the most simple and famous epistemologies of archaeology, with a strong bearing here, is Hawkes's 'ladder of inference' (Hawkes 1954). Hawkes's basic proposition was that, for those periods before the emergence of writing, 'material techniques are easy to infer to, subsistence-economics fairly easy, communal organisation harder, and spiritual life hardest of all [...] So the result appears to be that the more specifically human are men's activities, the harder they are to infer by this sort of archaeology'. Accordingly, it would seem that texts are more apt to inform about these 'difficult' dimensions, hardly graspable for prehistoric periods. This was just a paragraph in a longer paper concerned with archaeological theory and method but, as Robb (1998:330) puts it, 'Hawkes's dictum was essentially a formalization of common sense, and its intuitive appeal has helped to enshrine it in archaeological theory'.

While this view is inherent in several subsequent studies, Hawkes has increasingly become the object of criticism on several grounds. His opponents are not only those who advocate a more respectable status for archaeological materials in the 'artefacts *vs* texts' debate (for example, Andrén and Moreland, above) but also, more generally, archaeologists involved in the reconstruction of symbolic systems (for instance, Hodder 1982:11–12; Robb 1998:330–31), or even archaeometrists who, in spite of their obvious focus on materials and technologies, realise the inextricable cultural dimension of technology and, thus, the role of technological studies as a path to past social structures and beliefs (for example, Andrews and Doonan 2003:77–78). In an enlightening paper, Evans (1998) has complained that Hawkes has been turned into a sort of 'straw man', usually without noticing, first, that he was talking only about prehistoric archaeology – that is, an archaeology not subordinated to texts but simply devoid of them; and, second, that the four degrees of difficulty were never described as four watertight levels of hierarchical importance.

Only one of the case studies presented in this book explicitly refers to Hawkes's ladder. Graves-Brown thoroughly reviews the archaeological and textual evidence for the ritual and symbolic use of flint artefacts in Ancient Egypt. Using metaphor theory, she then explores the differences between both types of sources and, importantly, how this encounter changes over time. The resulting essay is informative with regard to Ancient Egyptian religion and to the past perception and use of different materials and sources, but it may also be inspiring as a methodological example for future studies in other areas. Clearly, Graves-Browns sides up with those who cite Hawkes to then contend that archaeology can indeed help access the 'hardest of all' arenas of past societies. In so doing, she also demonstrates the permeability of boundaries between technology and religion.

Freestone (this volume) takes a holistic approach to Roman glass-making, combining a review of relevant excerpts from Pliny's *Historia Naturalis* with archaeological, ethnographic, and analytical information. This allows him to decipher the written source and complete crucial bits of information regarding the raw materials employed and the organisation of glass production. Besides providing a meticulous understanding of how glass was made, this work offers a broad picture of the social and economic dimensions of glass manufacture, involving relationships between primary and secondary production workshops, as well as a wide international trade network to obtain soda, sand, and different additives. Even if not mentioning the famous 'ladder of inference', Freestone's final discussion makes a statement that reminds of Hawkes's original thought: 'Ideology and belief are areas which are in general far more easily conveyed through words than through the analysis of objects or structures. However, technology is an area where textual evidence may be quite limited' (Freestone, this volume, p. 97). Again, this acknowledgment compels neither subordinating archaeology to history nor constraining archaeological research to text-free zones. Freestone's study itself provides a proof of this.

The power of archaeological studies for the understanding of past societies – including those peoples and facets of their lives for which we have written records – is further reinforced when one considers the crucial role of material culture as a medium through which humans structure and express their position in the world. I come back to this idea after reviewing the different methodological approaches engaging materials and texts.

Methodological Approaches

Three different – if overlapping – methodological approaches can be identified in this area of interdisciplinary work, all of which are exemplified in this book. These may be termed, respectively, *problem-oriented*, *text-oriented*, and *material-oriented*. Although the first one seems more in line with conventional research practice, it may be argued that they are all necessary and complement one another.

Problem-Oriented Research

A usual touchstone for good practice in research is the existence of a well-formulated question justifying the work. Funding bodies, researchers, and readers often ask: What is the point of doing this? What are the expected results? In historical sciences, the question can usually be phrased as: What will this research tell me about past peoples? A problem is first posed, and, from this starting point, resolute scholars use any tool at hand to obtain the most accurate answer, which will often require combining archaeological, historical, and analytical sources.

The criticism or oblivion levelled against many early archaeometric studies stemmed precisely from their lack of a problem. Archaeological materials were analysed for disparate reasons, few of which had to do with past peoples or wider scholarly concerns. As a result, many of these early publications, although valuable as pioneer efforts, could now be equated to the characters of Pirandello's comedy – they are 'six thousand numbers in search of an archaeological problem'. On the contrary, few present-day doctoral students will get away without a clear statement of their 'research questions'.

Ashton's study (this volume) departs from a neat question, with clear historical and archaeological implications: Where was the main production centre of artistic material culture for Ptolemaic and Roman Egypt? Traditional research, mostly by Classicists, drew on the amount and quality of literary and historical works produced in Alexandria and tended to characterise this city as the most important artistic centre. Ashton reconsiders this scenario by critically reviewing the historical information and supplementing this with a wealth of largely ignored archaeological evidence. The answer to the question is compelling, as production remains of faience, stone sculpture, and plaster moulds for figurines strongly suggest that the source of Greek artistic innovation in this period was Memphite rather than Alexandrian.

This is a suitable example of interdisciplinary strategies integrated to address a research question, as well as an expression of the fruitful use of material remains from old excavations and long stored in an archaeological museum. However, as stated above, there are other valid approaches to interdisciplinarity.

Text-Oriented Research

It would be naïve to neglect that, in some instances, texts deserve centre stage even if material remains are to be considered too. This is not necessarily an adverse bias but may be simply a reflection of the complex nature of a given written source and/or its social or academic notoriety, which render it worthy of focused attention. A paradigmatic example of this type of text is the Bible, and another one, discussed in this book, is the body of cuneifrom-inscribed clay tablets often referred to as the 'Library of Assurbanipal'.

Using texts referring to glass as a case in point, Shortland (this volume) endeavours to unveil their intricate origins, purpose, and meaning, as well as their infor-mative potential. Like other texts addressed in this volume (see below), the cuneiform tablets present, to start with, the challenge of deciphering terms for which we lack a direct modern language counterpart but that may be better discernible as actual materials. Furthermore, the tablets were written in the 1st millennium B.C.E., but they are understood to constitute copies of 2nd-millennium B.C.E. originals – can we assume that nothing was changed or added to their content in 1,000 years of transmission? If they are to be read as mere technical recipes for the manufacture of glass, then we might expect copiers to have updated the information to adapt it to current knowledge. If, perhaps conversely, their role was primarily ritual and religious, maybe preserving their original form would have been more important. By showing that some parts of the recipes, if adequately followed, do lead to the production of glass, Shortland attests to the validity of this source for present researchers of ancient technologies. This is corroborated by the primary glass remains from Amarna, whose identification sheds light on, and is enlightened by, the cuneiform recipes. At the same time, the presence of symbolic elements in the text, and even the 'ritualisation' of a seemingly unnecessarily long glass production sequence, point toward social dimensions that reach beyond the realm of simple and utilitarian techniques. How can we interpret this? According to Shortland, we must consider the life-history of the text to understand its changing functions throughout the period. In this sense, just as these tablets may be illustrative of 2nd-millennium B.C.E. glass-making practice, they may also inform about much later – 1st millennium B.C.E. – lexicographic and literary enterprises.

In other words, only once we know the 'history of becoming' of the text can we feel confident to use it for the appropriate historical or archaeological research topics. The text exegesis itself reveals interesting information about the people who inspired and were inspired by the text. Subsequently, relevant information may be extracted for further research purposes. Thus we can see the *text-oriented* study as a necessary means but also as an end in itself. Similar views can be held on *material-oriented* studies.

Material-Oriented Research

Despite the encouraging advances in surveying techniques and landscape archaeology, excavation arguably remains the most important archaeological strategy to obtain raw data – materials and contexts – for the understanding of past societies. Archaeologists increasingly begin to dig having a fair idea of what they are going to unveil, which is of great assistance for the planning of specialist consultancy, finds processing, and conservation, but not refined enough to replace excavation or, at least, to avoid finding the unexpected. Virtually any field report can furnish examples of features, materials, or associations that cannot be immediately interpreted. The obvious question emerges – 'what is this?' – and *material-oriented* research is what often ensues.

Archaeometallurgical remains are one of those categories usually requiring devoted attention before they can be suitably identified and used for further research. Although it may be generally unproblematic to attribute slag or crucibles alike to the hodgepodge category of 'metallurgy', this assertion is as relatively bland as ending a report by stating 'this is a papyrus document' or, for the case above, 'this is indeed a cuneiform-inscribed tablet'. We need to go further – and archaeometric techniques often offer the best path.

Identifying specific raw materials and techniques is often a key to investigating cultural traditions and to understanding the social embeddedness of metallurgical practices. The availability and supply of different resources, the efficiency of craftspeople, and the sophistication of different technical strategies are but some of the issues with clear wider implications and providing pathways to ancient technologies and societies.

In her chapter, Bayley summarises the results of her extensive research on medieval remains of noble metal assaying and refining. The economic and cultural value of gold and silver in the Middle Ages underpins the importance of identifying the locations where these metals were processed, as well as the specific tools and techniques employed. Their degree of specialisation, and the fact that noble metals are only rarely left in archaeological workshops, mean that the task of identifying and interpreting such remains is arduous. A good example is provided by the vessels used for parting gold from silver by cementation, which Bayley identifies on the basis of their bleached fabrics and the presence of haematite on their inner surfaces. While this requires detailed scientific analyses and a modern understanding of the chemistry involved, medieval treatises furnish a necessary key for this ascription, given that such a parting method is no longer in use.

The chapter on brass production (Rehren and Martinón-Torres) is also based on a project that started with several cases of *material-oriented* research but later expanded onto a wider exploration. The main object of the study is the contrast between the practical side of brassmaking, as observed through the analysis of archaeological remains, and the perception and understanding of the process as described in a variety of written sources. It is argued that, although the technique employed for the production of brass remains essentially unchanged throughout the Middle Ages and the Renaissance, the understanding of the process evolves substantially, from early descriptions of how to 'colour copper yellow by art' to the later awareness of the presence of zinc and its contribution to the material properties of the alloy. The integrated approach to brass demonstrates that Renaissance rationalism and observations did not bring immediate changes in the metallurgical practice: Theory and practice evolved at different paces, sometimes seemingly unconnected. In this sense, this research provides an insight into a peculiar relationship between craft and science, while it helps assess the influence of the Renaissance mindset in a particular area of human activity. Crucially, this would not have been possible without the previous analytical study of several material assemblages that, aided by textual information, had been identified as brassmaking remains.

Materiality and Archaeometric Analyses

In line with some of the aforementioned traditional approaches to the texts and artefacts debate, Cartledge (1979:9; cited in Lloyd 1986:43) once noted: 'From archaeological evidence alone we may infer (relatively) much about material techniques, a considerable amount about patterns of subsistence and utilization of the environment, far less about social and political events, and least of all about mental structures, religious and other "spiritual" ideas and beliefs'. As with Hawkes, several scholars have been swift to respond with instances that clearly illustrate the contribution of archaeological studies to the reconstruction of political and ideological structures (for example, Lloyd 1986), and further examples of such attitudes are manifest in this book.

As a matter of fact, a stronger contention may be made: I believe that *any* contextual study of material culture, archaeological or not, should allow inferences of social, political, and mental structures. More specifically, there is a growing body of literature persuasively arguing that technological practice is inextricably embedded in context-specific social relationships, dynamics of knowledge transmission, and processes of construction and contestation of power and identity (for instance, Dobres 2000; Dobres and Hoffman 1999; Lemonnier 1993; Pfaffenberger 1992; Sillar and Tite 2000). Materials, products, and agents are placed within specific historical conditions and value systems, and these dimensions are analytically indivisible. Human behaviour entangles components that are often heuristically separated, such as technology, subsistence, society, and culture. Material culture – obviously including books, but not exclusively – is intrinsic to all. The human body and material culture are the most prominent media for social and cultural production and interaction, hence the study of materials is essential for the understanding of human behaviour. If this is the case even for present-day societies, it is more so for archaeological ones, for which the humans themselves are often no longer available as a study object. In short, I am keen to accept that archaeology is particularly suited to study technology but also to acknowledge that materials and technologies are products and expressions of social and cultural phenomena, and, as such, they inform about their societies and cultures of reference.

This realisation should suffice to make a case not just for the pertinence, but even for the need, of archaeological research of text-informed periods. But this can be taken further to argue that, in most cases, the very same principles used to defend archaeology can be used to defend archaeometry. That is to say, what justifies the 'destruction' of archaeological contexts via excavation should also hold up the investment in – and notional 'damage' by – scientific analyses of materials (see Rehren 2002). By and large, archaeological excavation is understood as a necessary means to recover information about past societies. Its destructive component may be lamented on as unavoidable, but its pertinence is rarely questioned insofar as 'knowledge' is to be obtained. Following the same line of reasoning, we have an obligation to exploit the informative potential of materials as far as possible, even if this requires a further stage of disturbance to the archaeological heritage – one of a minimal magnitude when compared to that of the excavation. As Rehren (2002:11) has argued, 'once damage to a site has been done, it is only a matter of courtesy to the site, and responsibility to the past, to maximise the knowledge gained in compensation'. Obviously, just as excavation, invasive sampling should be carefully evaluated and placed within wider research agendas. The worth of archaeometry is directly proportional to its relevance to archaeology as a whole.

The last decades have witnessed an encouraging development of theoretical frameworks aimed at encapsulating the interests and abilities of social and materials scientists. The *chaîne opératoire* is one of the most prominent such approaches, focusing on the reconstruction and understanding of technical sequences but increasingly placing emphasis on the social embeddedness of materials and humans (for example, Dobres 1999). By studying and interpreting technical information within the relevant historical and archaeological contexts, archaeometrists are contributing to the understanding of past social and mental structures, transmissions of knowledge, relationships between humans and the environment, perceptions of materials, and identities – to name but a few of the aspects that interest historians, archaeologists, and anthropologists alike. Within this framework, the identification of 'technological choices' has proved an inspiring challenge for materials scientists, who can reveal otherwise inaccessible, subtle selections in the manufacture and use of materials, as a necessary foundation to then investigate the broader cultural and environmental reasons leading to those choices (Lemonnier 1993; Schiffer 2003; Schiffer et al. 2001; Sillar 2003; Sillar and Tite 2000; Tite, Kilikoglou, and Vekinis 2001). Several studies in this book are more or less implicitly inscribed in these trends, namely, reconstrucing *chaînes* (Freestone, Shortland, Rehren, and Martinón-Torres) or focusing on more specific choices or technical steps (Bayley, Jackson and Smedley, Pérez-Arantegui and Pardos) within their wider sociocultural contexts. Recently, there have been calls to recognise the 'materiality' of material culture, emphasising the role of sensorial and material dimensions of things in the way they are perceived and engaged in human life and advocating the inextricable engagement between materials and ideas (DeMarrais, Gosden, and Renfrew 2004; Jones 2002, 2004; Meskell 2005; Toren 1999). Again, materials specialists in archaeometry are ideally suited to contribute their expertise about the physical, chemical, and mechanical properties of past materials and try to envisage how these affected their perception, symbolisation, and use.Some of these issues are illustrated in the discussion of the different perceptions of brass and brassmaking in medieval and later times (Rehren and Martinón-Torres, this volume).

In short, the scientific analysis of archaeological materials may be substantiated on the grounds of 'appropriateness' (Rehren 2002) and provided that there is 'a common theoretical umbrella' (De Atley and Bishop 1991:367) enabling communication among different specialists who often address similar problems with dissimilar sources and methodologies. It is important to emphasise, moreover, that a substantial number of analytical studies make good use of archaeological materials that would otherwise remain ignored in museum stores – when not abandoned in spoil heaps. Archaeometric approaches repeatedly show that behind the appeal (or lack therein) of ancient materials lies much more than a craftsperson's dexterity. Furthermore, instrumental analyses often reveal the clues to disentangle the paradoxes that are sometimes apparent when materials and texts are confronted.

As noted above, many archaeometric studies draw on data from written documents, namely, technical treatises. However, in a survey of reviews and compilations of the relationships between artefacts and texts (Andrén 1998; Beaudry 1988; Bintliff and Gaffney 1986; De Cunzo and Jameson, 2005; Ferguson 1977; Funary, Jones, and Jones 1999; Hall and Silliman 2006; Moreland 2001; Orser 2002; Orser and Fagan 1995; Spencer-Wood 1987), neither this type of book nor science-based case studies are addressed to any significant degree. When the information from technical treatises is explicitly confronted to that obtained via archaeological and archaeometric analyses, a number of observations emerge.

Technical Texts and Technical Studies: Some Potentials and Risks

With one of the earliest examples in the cuneiform tablets, a plethora of written texts have historically collected recipes for the production of textiles, glass, food, dyes, ceramics, metals, medicines, and potential solutions to many other 'technical problems'. Because these often appear written in a formal and systematic style, we have a natural tendency to assume that they are accurate, comprehensive, and unbiased accounts of how things are or were done. The first obvious point to be emphasised is that texts are written by subjects and are, as such, charged with subjectivity.

Even if recorded with encyclopaedic aims, or merely as memory aids, written documents have gone through the sieves of humans who observed, experienced, transmitted, copied, and documented whatever is being described or depicted. Few self-respecting scholars will present a historical reconstruction without a rigorous critical approach to the documentary evidence, just as archaeologists are expected to understand site formation processes before interpreting stratigraphy and material remains. This is one of the reasons justifying the text-oriented and material-oriented research approaches discussed above. However, an explicit critique of the sources – materials or texts – is not always present when researchers deal with apparently straightforward technical texts, or their material counterparts.

Past historiography is copious of examples of historical reconstructions that were exposed when their sources were revealed as not such accurate depictions of reality as originally thought. In this sense, it may be the historian's job, and not that of the archaeologist, to investigate the origins, *raison d'etre,* and underlying implications of a text, but archaeometric research can provide crucial information to this end. This also applies to descriptions of technical processes. An example of this collaboration was discussed above with regard to the role and the meaning of the cuneiform tablets (Shortland, this volume). Another mesmerising example appears in the study of hispano-moresque lustre pottery by Pérez-Arantegui and Pardos (this volume). Understanding the technology of production of the metallic lustre exhibited by some medieval Islamic ceramics has required intensive analytical and experimental work, not least because the recipes were zealously guarded, and hence very few extant documents describe the manufacturing process. One of the few available texts was collected by the infamous Dutchman Henrique Cock, a Renaissance intellectual and member of King Phillip II's Royal Guard, as heard from a potter in Muel (Spain). Among the various ingredients detailed, the potter mentioned 'a little copper', as duly collected in Cock's account. What Cock probably did not know, however, was the crucial importance of this inaccurate 'little copper', as the precise amount of copper added was the key to the different colours and optical effects achieved in lustre decoration. As it seems, the cunning potter satisfied the request of this member of the Royal Guard but was deliberately vague so that his guild's secret was not given away in full. It is only thanks to the comparative analyses of lustre ceramics from Muel that we can infer the witty subtlety of the text.

In other cases, the information contained in a text may not be intentionally deceitful, but researchers must be aware of their possible biases and imprecisions. Pliny's *Historia Naturalis,* to name a relevant example, constituted an unparalleled effort of erudition in exhaustively collecting information drawn from other writers together with his own observations and thoughts. However, this vast work is not devoid of contradictions, inaccuracies, and errors (Healy 1999; Freestone, this volume). In the Middle Ages, a corpus of technical texts appears, many copying – and some augmenting – the information from Classical documents. Just as these treatises vary in their quality and detail, so does their potential as historical sources. In her study of medieval precious metal refining, Bayley (this volume) decidedly privileges Theophilus's *De diversis artibus* (Hawthorne and Smith 1979) as a primary source of information, over the also famous *Mappae Clavicula* (Smith and Hawthorne 1974). The reason is that, whereas the author of the latter simply compiled information and clearly lacked an understanding of many of the processes he was describing, Theophilus was a practitioner writing for other craftsmen and hence offers more comprehensive and apparently accurate descriptions. However, there are problems even when working with technologically informed texts. For instance, as noted by Bayley, authors often give alternative recipes or solutions to technical problems, and without archaeological work it is virtually impossible to document which ones may have been more established, and why.

The opposite problem may occur as well: When only one written description of a process is available, we are at risk of accepting that this was the 'standard' way of doing things, whereas in fact it may just represent a given author's view or own experience. In these cases, also the study of archaeological remains helps diversify the picture by illustrating different techniques or practices employed in the solution of similar technical problems, as a starting point to then address the possible reasons behind these technological choices. In some instances, these will be a reflection of the versatility of technological practices, which often have to 'make do' and improvise *ad hoc*, cost-effective solutions that differ from those suggested in the manual. In other cases, different traditions or social and environmental constraints may be at stake. A case in point is that presented by Jackson and Smedley (this volume). Because of Theophilus's account, it is widely accepted that beech wood ash was the primary source of alkali for glassmaking in medieval Northern Europe. By thoroughly complementing the textual record with archaeological, vegetational, analytical, and experimental data, Jackson and Smedley question the widespread use of this species and argue that this was but *one* of the various wood ashes used.

There is a further characteristic of historical treatises that highlights the need of complementary archaeological and scientific approaches. Technological practices involve the provision, processing, and utilisation of materials, but the perception and categorisation of different substances are contextually determined. In this sense, an ingredient mentioned in a written source may be, to start with, difficult to translate, but even when translated it may prove hopeless to figure out what is signified by the term. Many explanations may be proposed: The term may be part of jargon that we are no longer acquainted with, its meaning may have evolved to denote something different, or it may simply refer to a substance that we are not familiar with or that we categorise in a completely different way. Examples are abundant. In Amarna, the discovery of a Late Bronze Age glass frit that could be technically interpreted as 'intermediate primary glass', still requiring remelting and homogenisation, reinforces one of the proposed translations for the term '*zukû*-glass' of the cuneiform texts (Shortland, this volume). In a similar vein, Freestone (this volume) proposes material equivalents for '*magnes lapis*' and 'lustrous stones', mentioned by Pliny as ingredients for glassmaking. The arguments are based on extensive analyses of Roman glass and a careful consideration of glass chemistry, which allow a degree of precision that would not be achievable using linguistics alone. Also Smith and Hawthorne (1974:16) discuss several medieval terms that 'occur in such diverse chemical contexts that identification with a single substance would be quite misleading'.

It was indeed the eagerness to clarify the historically corrupt terminology in mineralogy and metallurgy that inspired Georgius Agricola to write his sizeable treatise *De re metallica*, published in 1556 (Hoover and Hoover 1950). As a humanist and Renaissance man, Agricola was concerned with re-establishing the appropriate Latin denominations. In addition, he was moved by a desire to raise the public profile of the mining and metallurgy profession, hoping to attract potential capitalist investors. Agricola's tome, together with Vannoccio Biringuccio's *De la pirotechnia* (1540; Smith and Gnudi 1990) and Lazarus Ercker's *Beschreibung allerfürnemisten mineralischen Ertzt- und Bergktwercks Arten* (1580; Sisco and Smith 1951) epitomise for the realm of metallurgy a general momentum in the production of technical literature. This acme was supported by the spread of the printing press and accompanied by a vehement – often institutionally sponsored – humanist interest in recording knowledge that was the beginning of our academic keenness on publishing. This surge of technical literature led to what Long calls an 'openness of knowledge' (Long 1991, 2001), which enabled craftspeople and laymen alike to learn and apply otherwise elusive skills and recipes. Undeniably, the printing press and the widespread distribution of these books may have contributed to the standardisation of certain metallurgical practices (Eisenstein 1979). However, before taking for granted the impact of books on the technological arena, it is necessary to assess the literacy of the assumed readership (Driscoll 1988; Moreland 1991) – significantly,

Agricola's book was written in Latin to ensure a learned audience, which would have left out most ordinary craftspeople. Furthermore, a text could be copied for generations without incorporating practical changes taking place across space and through time. For example, the furnaces and assaying equipment described by Ercker in the 16th century appear virtually unchanged in 18th-century treatises (Martinón-Torres and Rehren 2005); was the technology so conservative for 200 years, or is this simply the result of one author copying from another? Archaeology may have an answer to this. Conversely, we may have a long hiatus from the moment an innovation is theoretically conceived – and printed in books – until it is incorporated in routine technical practice (Hodges 1972; Rehren and Martinón-Torres, this volume).

There is one further critical problem when using technical treatises to study past technologies, and this is expressively encapsulated by Freestone (this volume, p. 97): 'the details of an operational sequence are extremely difficult to commit to text, as anyone who has attempted to follow the instructions included with a piece of modern, flat-packed furniture will testify'. Technical processes are generally difficult to put down in words, not least because often this requires a rationalisation and verbalisation of procedures that would not normally take place.

Apprenticeship, as the main channel for the transmission of specialised technical knowledge, usually had little to do with words but rather with practice (Coy 1989). Among others, Singleton's study of Japanese pottery-making apprenticeship illustrates this: 'When an apprentice presumes to ask the master a question, he will be asked why he has not been watching the potter at work, where the answer would be obvious. [. . .] A master is simply not expected to teach craft skills in didactic instruction. It is as improper to ask that he do so, as it is considered ineffective in transmitting the craft skills' (Singleton 1989:26). Technology is the realm of 'learning by doing'. Furthermore, technical processes are often carried out in a particular way because that is simply 'the way things are done' or, perhaps, because that is the way things have *always* been done (Lemonnier 1986:165). Although not desdaining the active decisions made by individual and social agents (Dobres 2000), or advocating an excessive 'technological somnambulism' (Pfaffenberger 1988), one cannot fail to notice that technical activities involve routine and systematic repetition, usually without questioning the appropriateness of the methods followed. It is then difficult to explain what one does because nonverbal, learned, apprehended factors lead the action. As aptly put by Pfaffenberger (1992:508), 'the portion of technical knowledge that people can verbalize represents only the tip of the iceberg'; owing to this, written treatises alone will never explain technological choices in full. The 'ritualisation' of technical activities, broadly understood as the unquestioned repetition of a procedure, may be constrained by religious taboos or simply sanctioned by tradition, but either way it has two clear consequences: First, it severely constrains innovation and change; second, it makes the verbal description and explanation of these processes increasingly difficult. Both of these factors are at stake in Bronze Age glassmaking (Shortland, this volume), just as they are in medieval and later brassmaking (Rehren and Martinón-Torres, this volume).

A somewhat different set of technical documents is that used by Badcock and Crossley (this volume), although some of their potentials and problems are common to those discussed above. Their main body of documentary information are the maps, notebooks, and correspondence from the Fairbanks, a family of surveyors who worked intensively in Sheffield between 1750 and 1850. This exceptional source of information keeps facilitating the job of archaeologists working in brownfield sites, since it often offers a preliminary insight into the presence of industrial sites and their distribution. This is particularly valuable within the usual practical constraints of rescue archaeology, because it allows a more precise planning of targeted excavation and conservation strategies. In addition, the documents provide details about landholding and business relationships that would be hardly graspable from archaeological data alone. The truthfulness of historical maps, however, cannot be taken for granted, and indeed archaeological work enables Badcock and Crossley to verify on the ground the degrees of cartographic precision achieved through the application of navigation and chart-making skills. Furthermore, maps are seldom concerned with the internal distribution of large industrial buildings, hence leaving out relevant information for those interested in our recent industrial past. One captivating example is shown in their work on the Jessop's Brightside steelworks, where archaeological excavations revealed that, in spite of the large outer dimensions of the workshop, this was internally subdivided in very small working units, each with its own hearth. Such an insight into the organisation of production shows that the growth in scale was not accompanied by a substantial change in the technology – perhaps a further indication of the usually slow pace of technological change. As Badcock and Crossley note, comparable surveyors' archives exist for other areas, and their own work should set a standard for their thorough study and exploitation.

Looking Forward

Both materials and texts are cultural expressions. At a general level, however, different conditionings tend to affect their respective generation and use, as well as their prospects of preservation and recovery. From this, it follows that they may inform about different – if complementary – human dimensions and that their combined use as sources for the study of the past will furnish more comprehensive and balanced pictures. Nonetheless, it would be futile, and counterproductive, to try to demarcate very precisely which subject areas appear more suited for either archaeology- or text-based research. While the case studies reviewed here illustrate the worth of multidisciplinary approaches, the somewhat diverging reflections on their sources presented by several authors indicate that, in many cases, their epistemological inferences may be applicable only to their specific case studies.

Is there any hope, then? Does any pattern emerge? What do we learn from this review? The most obvious lesson is the need of a critical approach to any source, be it material or textual. Although it may appear as a tautology, this is not always considered with enough rigour. It is us, researchers, who mediate to turn materials and texts into sources of information about past peoples (Galloway 2006). And we can do so only by investigating the *who*, the *what for*, the *whom for*, and the *why* behind the entities we are confronted with. These dimensions often linger between the conscious and the unconscious, the individual and the social, and it is our primary duty to clarify them.

In this sense, I propose looking at documents and artefacts together and in their contexts, rather than accepting universal assumptions or excessive academic specialisations. It is increasingly clear that historical and archaeological research should focus on the reconstruction of contexts, which can then be interpreted by analogy or association to reference frameworks. Texts, materials, and scientific data allow the reconstruction of richer contexts that, subsequently, enable the identification of patterns, processes, and specificities. When an ambiguity or poor fit is identified between our sources, a careful consideration of their respective formation processes and dynamics usually helps us to reconcile them beyond merely assigning any discrepancy to the realm of the exception (Leone and Crosby 1987).

These issues have been argued from different epistemological perspectives by Andrén (1998) and Moreland (2001), but we still find them rarely applied to focused case studies such as those presented in this book. As pointed out above, even text-oriented and material-oriented approaches benefit from resorting to each other, and they require cautious attention to the context for the material, for the text, and, crucially, for their encounter (Andrén 1998). The variable nature of the encounters between materials and documents is what makes every case study distinct and particular, while serving as a foundation for the discussion of wider issues, such as literacy, the perception and understanding of different materials, agency, and the social dimensions of material culture – including texts.

An explicit focus on these encounters provides a useful methodological strategy to address interesting questions about human societies that will hopefully engage historians, archaeologists, and scientists alike. Particularly, I urge researchers of past societies to go beyond the all-too-easy 'what did they do?' to explore the 'why did they do what they did?' and even 'why did they write what they wrote about what they did?' Answering these questions requires a detailed consideration of specific contexts and the critical and integrated use of textual, archaeological, and scientific approaches. In turn, this opens a fascinating path for the understanding of humans and their choices, and a challenge for comparative approaches to individuals and societies, both past and present.

Encouraging as it may seem, this proposal requires some practical commitments. Namely, the present-day compartmentalisation of academia may get in the way of a fluid communication between specialists. I have already argued that engaging questions within a broad overarching paradigm – rather than interdisciplinarity as a matter of principle – should promote more dialogue and cooperation. In addition, in some cases, it may still be necessary to make an effort to present the results of our research to target audiences who may not be familiar with all the relevant jargon, methodologies, and specialised publications. Recently, the directors of the research project at La Isabela, the first European town in America, have set a standard by producing two monographs – one primarily aimed at archaeologists (Deagan and Cruxent 2002a), the other one at non-archaeologists (Deagan and Cruxent 2002b). Archaeological and historical information is entrenched in both volumes, and surely many readers will benefit from both. The objective of such an enterprise was not to deepen the divides among specialists but rather, to facilitate communication. As the authors explain, they tried to accomplish 'both the presentation of technical procedures and data that can be used for comparative archaeological analysis with other sites and, at the same time, the presentation of a coherent, integrated story about the site and its significance that is useful to researchers who are not necessarily archaeologists' (Deagan and Cruxent 2002a:xxii). It is further encouraging to see that the results of ongoing scientific analyses of material culture from La Isabela are also published in such an inclusive way (Thibodeau et al. 2007).

Arguably, archaeometrists face the biggest challenge if they are to make their work available and useful for a wide spectrum of scholars. On the one hand, the technical nature of their methods and data makes it particularly difficult for nonspecialists to extract relevant information. On the other hand, scientists are often constrained by official research assessment criteria, meaning that only publications in certain journals will be accounted for in their professional evaluations. While we await more balanced ways of assessing academic merit and output, one can still find grounds to encourage archaeometrists to digest their own data and present it to a broader readership. First, it will always be more productive – and rewarding! – to find the results of their work enjoyed, used, and built on by a more eclectic range of scholars and laypeople alike. Second, this is a matter of professional responsibility: If we are to continue using public heritage (and often, financial resources) for our research, this can be justified only by making our results relevant and comprehensible for as many people as possible. If this requires the extra effort of writing two publications – one more technical, the other more narrative – this seems a fair price to be paid.

We can now return to the departure point: Why should archaeologists, historians, and scientists take one another seriously? I believe that they should do so because, by combining their strengths, their ability to understand past societies is multiplied exponentially. Furthermore, sharing questions and interests, and dealing with them together, may prove as challenging as rewarding. We have come a long way, and promising times lie ahead of us.

References

Andrén, A. 1998. *Between Artifacts and Texts: Historical Archaeology in Global Perspective.* New York: Plenum Press.

Andrews, K., and R. Doonan. 2003. *Test Tubes and Trowels. Using Science in Archaeology.* Stroud: Tempus Publishing.

Beaudry, M. C. (ed.). 1988. *Documentary Archaeology in the New World.* Cambridge: Cambridge University Press.

Bintliff, J. L., and C. F. Gaffney (eds.). 1986. *Archaeology at the Interface: Studies in Archaeology's Relationships with History, Geography, Biology and Physical Science.* International Series 300. Oxford: BAR.

Cartledge, P. 1979. *Sparta and Lakonia: A Regional History, 1300–362 BC.* London: Routledge.

Champion, T. C. 1990. 'Medieval archaeology and the tyranny of the historical record', in D. Austin and L. Alcock (eds.), *From the Baltic to the Black Sea: Studies in Medieval Archaeology.* London: Routledge, 79–95.

Coy, M. W. (ed.). 1989. *Apprenticeship: From Theory to Method and Back Again.* New York: State University of New York Press.

De Atley, S. P., and R. L. Bishop. 1991. 'Toward an integrated interface for archaeology and archaeometry', in R. L. Bishop and F. W. Lange (eds.), *The Ceramic Legacy of Anna O. Shepard.* Niwot: University Press of Colorado, 358–81.

De Cunzo, L. A., and J. H. Jameson Jr (eds.). 2005. *Unlocking the Past: Celebrating Historical Archaeology in North America.* Gainesville: University Press of Florida.

Deagan, K., and J. M. Cruxent. 2002a. *Archaeology at La Isabela: America's First European Town.* New Haven, CT: Yale University Press.

Deagan, K., and J. M. Cruxent. 2002b. *Columbus's Outpost among the Taínos. Spain and America at La Isabela, 1493–1498.* New Haven, CT: Yale University Press.

DeMarrais, E., C. Gosden, and C. Renfrew (eds.). 2004. *Rethinking Materiality: The Engagement of the Mind with the Material World.* Cambridge: McDonald Institute for Archaeological Research.

Dobres, M.-A. 1999. 'Technology's links and *chaînes*: The processual unfolding of technique and technician', in M.-A. Dobres and C. R. Hoffman (eds.), *The Social Dynamics of Technology: Practice, Politics and World Views.* Washington: Smithsonian Institute Press, 124–146.

———. 2000. *Technology and Social Agency: Outlining a Practice Frame-work for Archaeology.* London: Blackwell.

Dobres, M.-A., and C. R. Hoffman (eds.). 1999. *The Social Dynamics of Technology: Practice, Politics, and World Views.* Washington, DC: Smithsonian Institution Press.

Driscoll, S. T. 1988. 'The relationship between history and archaeology: Artefacts, documents and power', in S. T. Driscoll and M. R. Nieke (eds.), *Power and Politics in Early Medieval Britain and Ireland.* Edinburgh: Edinburgh University Press, 162–87.

Eisenstein, E. L. 1979. *The Printing Press as an Agent of Change: Communications and Cultural Transformations in Early-Modern Europe.* Cambridge: Cambridge University Press.

Evans, C. 1998. 'Historicism, chronology and straw men: Situating Hawkes' "Ladder of inference"', *Antiquity* 72:398–404.

Ferguson, L. (ed.). 1977. *Historical Archaeology and the Importance of Material Things.* Lansing, Mich.: The Society for Historical Archaeology.

Funary, P. P., M. Jones, and S. Jones (eds.). 1999. *Historical Archaeology: Back from the Edge.* One World Archaeology 31. London: Routledge.

Galloway, P. 2006. 'Material culture and text: Exploring spaces within and between', in M. Hall and S. W. Silliman (eds.), *Historical Archaeology.* Malden, MA: Blackwell Publishing, 42–64.

Hall, M., and S. W. Silliman (eds.). 2006. *Historical Archaeology.* Malden, MA: Blackwell Publishing.

Hawkes, C. 1954. 'Archaeological method and theory: Some suggestions from the Old World', *American Anthropologist* 56:155–68.

Hawthorne, J. G., and C. S. Smith. 1979 [1963]. *Theophilus: On Divers Arts. The Foremost Medieval Treatise on Painting, Glassmaking and Metalwork.* New York: Dover Publications Inc.

Healy, J. F. 1999. *Pliny the Elder on Science and Technology.* Oxford: Oxford University Press.

Hodder, I. 1982. 'Theoretical archaeology: A reactionary view', in I. Hodder (ed.), *Symbolic and Structural Archaeology.* Cambridge: Cambridge University Press, 1–16.

Hodges, H. W. M. 1972. 'Archaeology and the history of mediaeval technology', in D. E. Strong (ed.), *Archaeological Theory and Practice.* London: Seminar Press, 265–73.

Hoover, H. C., and H. L. Hoover. 1950 [1912]. *Georgius Agricola: De Re Metallica. Translated from the First Latin Edition of 1556.* New York: Dover Publications Inc.

Jones, A. 2002. *Archaeological Theory and Scientific Practice.* Topics in Contemporary Archaeology. Cambridge: Cambridge University Press.

Jones, A. 2004. 'Archaeometry and materiality: Materials-based analysis in theory and practice', *Archaeometry* 46:327–38.

Lemonnier, P. 1986. 'The study of material culture today: Towards an anthropology of technical systems', *Journal of Anthropological Archaeology* 5:147–86.

Lemonnier, P. (ed.). 1993. *Technological Choices. Transformation in Material Cultures since the Neolithic*. London: Routledge.

Leone, M. P., and C. A. Crosby. 1987. 'Middle-range theory in historical archaeology', in S. M. Spencer-Wood (ed.), *Consumer Choice in Historical Archaeology*. New York: Plenum Press, 397–410.

Little, B. J. (ed.). 1992. *Text-Aided Archaeology*. Boca Raton, FL: CRC Press.

Lloyd, J. A. 1986. 'Why should historians take archaeology seriously?', in J. L. Bintliff and C. F. Gaffney (eds.), *Archaeology at the Interface. Studies in Archaeology's Relationships with History, Geography, Biology and Physical Science* (BAR International Series, 300). Oxford: BAR, 40–51.

Long, P. O. 1991. 'The openness of knowledge: An ideal and its context in 16th-century writings on mining and metallurgy', *Technology and Culture* 32:318–55.

———. 2001. *Openness, Secrecy, Authorship: Technical Arts and the Culture of Knowledge from Antiquity to the Renaissance*. Baltimore: Johns Hopkins University Press.

Martinón-Torres, M., and Th. Rehren. 2005. 'Alchemy, chemistry and metallurgy in Renaissance Europe: A wider context for fire assay remains', *Historical Metallurgy* 39:14–31.

Meskell, L. (ed.). 2005. *Archaeologies of Materiality*. Oxford: Blackwell.

Moreland, J. 1991. 'Method and theory in medieval archaeology in the 1990s', *Archeologia Medievale* 18:7–42.

———. 2001. *Archaeology and Text*. London: Duckworth.

Orser Jr, C. E. (ed.). 2002. *Encyclopaedia of Historical Archaeology*. London: Routledge.

Orser Jr, C. E., and B. M. Fagan. 1995. *Historical Archaeology*. New York: Harper Collins College Publishers.

Pfaffenberger, B. 1988. 'Fetishised objects and humanised nature: Towards an anthropology of technology', *Man* 23:136–52.

———. 1992. 'Social anthropology of technology', *Annual Review of Anthropology* 21:491–516.

Rehren, Th. 2002. 'Reply to Tite: Invasive sampling – or why do we excavate?', *PIA, Papers from the Institute of Archaeology* 13:9–12.

Robb, J. E. 1998. 'The archaeology of symbols', *Annual Review of Anthropology* 27:329–46.

Schiffer, M. B. 2003. 'Comments II: Properties, performance characteristics and behavioural theory in the study of technology', *Archaeometry* 45:169–72.

Schiffer, M. B., J. M. Skibo, J. L. Griffiths, K. L. Hollenback, and W. A. Longacre. 2001. 'Behavioral archaeology and the study of technology', *American Antiquity* 66:729–38.

Sillar, B. 2003. 'Comments III: Technological choices and experimental archaeology', *Archaeometry* 45:173–81.

Sillar, B., and M. S. Tite. 2000. 'The challenge of "technological choices" for materials science approaches in archaeology', *Archaeometry* 42:2–20.

Singleton, J. 1989. 'Japanese folkcraft pottery apprenticeship: Cultural patterns of an educational institution', in M. W. Coy (ed.), *Apprenticeship: From Theory to Method and Back Again*. New York: State University of New York Press, 13–30.

Sisco, A. G., and C. S. Smith. 1951. *Lazarus Ercker's Treatise on Ores and Assaying, Translated from the German Edition of 1580.* Chicago: The University of Chicago Press.

Smith, C. S., and M. T. Gnudi. 1990 [1959]. *The Pirotechnia of Vannoccio Biringuccio: The Classic Sixteenth-Century Treatise on Metals and Metallurgy.* New York: Dover Publications Inc.

Smith, C. S., and J. G. Hawthorne. 1974. '*Mappae Clavicula*: A little key to the world of medieval techniques', *Transactions of the American Philosophical Society* (new series), 64(4).

Spencer-Wood, S. M. (ed.). 1987. *Consumer Choice in Historical Archaeology.* New York: Plenum Press.

Thibodeau, A. M., D. J. Killick, J. Ruiz, J. T. Chesley, K. Deagan, J. M. Cruxent, and W. Lyman. 2007. 'The strange case of the earliest silver extraction by European colonists in the New World', *Proceedings of the National Academy of Sciences* 104:3663–666.

Tite, M. S., V. Kilikoglou, G., and Vekinis, G. 2001. 'Strength, toughness and thermal shock resistance of ancient ceramics, and their influence on technological choice', *Archaeometry* 43:301–24.

Toren, C. 1999. *Mind, Materiality, and History.* London: Routledge.

Young, B. K. 1992. 'Text aided or text misled? Reflections on the uses of archaeology in medieval history', in B. J. Little (ed.), *Text-Aided Archaeology.* Boca Raton, FL: CRC Press. 135–47.

CHAPTER 2

Licking Knives and Stone Snakes: The Ideology of Flint Knives in Ancient Egypt

CAROLYN GRAVES-BROWN

ABSTRACT *Flint tools, the study of which is traditionally the preserve of the prehistorian, were used in Egypt throughout the Pharaonic Period. Here the lithic technology of Dynastic Egypt is employed to explore the relationship between archaeology and text, restricting discussion to selected aspects of flint knife manufacture and use following the introduction of writing (ca 3100 B.C.E.) to the Greco-Roman Period (ca 300 C.E.). The main focus is placed on ideological aspects where, following the Hawkesian ladder of inference (Hawkes 1954), one might assume that text would play the significant role. Through a straight comparison of evidence from each area, it is argued that archaeology can illuminate ideology, too, that there is overlap between the two. Finally, an attempt to explain difference, rather than simply using one tool as a handmaiden of the other, may shed light on ideological development and bring new areas into focus. Aspects of metaphor theory are used in an attempt to achieve this.*

Chronology

Predynastic	5500–3100 B.C.E.
Early Dynastic	3100–2686 B.C.E.
Old Kingdom	2686–2181 B.C.E.
First Intermediate Period	2181–2055 B.C.E.
Middle Kingdom	2055–1650 B.C.E.
Second Intermediate Period	1650–1550 B.C.E.
New Kingdom	1550–1069 B.C.E.

Third Intermediate Period	1069–747 B.C.E.
Late Period	747–332 B.C.E.
Ptolemaic Period	332–30 B.C.E.
Roman Period	30 B.C.E.–395 C.E.

Introduction

Both text and archaeology are incomplete or unclear evidence of past activity. For religion, text has traditionally been used as the primary form of evidence. For technological factors, archaeology is often given primacy. Here I show, in an examination of the ideology surrounding the technology of flint in ancient Egypt, that text and archaeology can be used together, not simply through searching for complementary evidence but also through exploring differences. Even though text is often privileged for explaining ancient religions, it has limitations. Although the nature of Egyptian texts vary from period to period and literary genre to literary genre, they generally relate to an ideal, archaic world, perhaps a world occupied only by a small elite (Gardiner 1961:54–60; Parkinson 2002:49, 66–78). Baines (1983:584) estimates that in most periods only 1% of the population were literate and sees little evidence for cultural heterogeneity between classes before the Late Period (Baines 1983:586). Magic/religious texts in particular used age to reinforce their status (Baines 1983: 588), hence their archaic nature. Where flint is mentioned in literature, it is usually within such religious texts. To study religion merely through text reduces it to a numinous irrelevance, since ideology depends for its power on the body in the real world. Archaeology, however, deals with the multivocal and subtle nature of the object, which is more akin to imprecise concepts of humanity acting within the world; in contrast, text tends to be specific and precise. Although archaeology concerns primarily praxis and therefore relates to the embedded relations of ideology, its discontinuous, subtle nature means we can only dimly perceive through it the specifics of doctrine and the poetic. Here, text may appear as a more suitable tool.

Technology is here taken to mean extraction and fashioning of materials, including ideological practice impinging on action. Like text and archaeology, technology and religion were traditionally considered incompatible. Yet, the embeddedness of technology is now generally recognised (Dobres 2000; Lemonnier 1986; Michael 2000; Pfaffenberger 1988, 1992; Walker 2001). Indeed, in Dynastic Egypt, religion and technology appear conjoined, with priests and kings participating in both. Technological activities depicted on tomb walls are part of the creative, ideological process necessary for rebirth in the afterlife. Words such as *msj* mean both 'to give birth' and also to 'create', as in a statue.

Traditionally, Egyptologists have considered lithics a part of technology, in the essentialist sense, rather than a part of religion. Exceptions include attempts to differentiate between the sacred and profane by examining whether material is found on domestic or temple and burial sites (Kromer 1978; Schmidt 1992; Spurrell 1891:51). In these studies, there is little recognition that the sacred and the profane may not have been considered quite so divisable in ancient Egypt, with clearly religious objects turning up on secular sites. Shirai (2005), however, using a purely archaeological approach for Predynastic Egypt, clearly recognises the difficulty of separating the sacred from the profane. Midant-Reynes (1981) looked at words for flint in ancient Egyptian and in doing so brought out the mythical use of the material. Using artefacts, Midant-Reynes (1987) explored possible symbolism behind Predynastic knives and in particular the duality apparent in the treatment of ripple flaked pieces. Aufrère (1983, 1991:vol. 2, 263–569), using primarily textual evidence, has looked at how flint was understood in terms of deities and cosmologies. Harris (1961:138–39, 228–29, 233) has briefly explored the textual ramifications of flint that bring out its use in religious life. Thus, exploration of the ideology of flint in ancient Egypt has tended to deal with the Early and Predynastic Periods. Where it considers Dynastic remains it mainly employs textual material. To counter this, here I intend to bring together literature, iconography, and archaeology to explore Dynastic material. It should be noted that flint knives were manufactured in Egypt until at least the New Kingdom – for example, at Memphis (Giddy 1999) and East Karnak (Miller 1985).

Text here means the written word, and archaeology the material object, and we recognise, of course, that artefacts with text are also archaeological and that objects can be studied through text and vice versa. I have opted for placing iconography with archaeology on the grounds that it features in nonliterate societies. However, in Egypt, writing had an iconographic quality and pictorial representations illuminated text.

Extracting textual evidence for flint knives is complicated. One of the words for flint, *ds*, is also a word for 'knife'. Some commentators believe the *ds* knife to have been conceptually flint and in religious literature a tool to punish wrongdoers. In some translations, therefore, *ds* is given as 'flint knife'. I have some sympathies with this but have restricted the assumption that a knife was flint to those instances where the stone determinative is used after the word, or where text is so early as to suggest that it is more likely that a flint rather than a metal knife was used. For example, following Faulkner (1969:54), I have assumed knives in the *Pyramid Texts* (ca 2350 B.C.E.) to be flint.

Archaeology-Centred Evidence

The following areas are those in which, I believe, archaeology best illustrates ritual use of flint: ritual butchery, protection, creational, non-utilitarian forms, and heirlooms. These areas are not mutually exclusive but separated here for pragmatic reasons. Each is discussed first from the viewpoint of archaeology, then iconography, and, last, textual evidence.

Ritual Butchery

How would we deduce flint use in *ritual* butchery? Wear analysis demonstrates only the physical act of cutting. As Tillmann (1992:197) has pointed out, archaeological examples of flint knives in slaughter contexts are rare. One exception comes from 1st-Dynasty Helwan (3100–2890 B.C.E.), where a flint knife was found in a magazine in a layer of ox bones and two other flint knives were found in a cross shape at the bottom of the structure (Saad 1951:10–11, Pl. VII–VIII).

Flint knives, at least until the end of the Old Kingdom, are more frequently found in burial and temple contexts (Schmidt 1992:87; Tillmann 1992) – sites usually classified as ritual. In contradiction, Kromer (1978:30) believes those from Giza are profane. Since the purpose of the Giza settlement seems to have been to build and to serve mortuary monuments, one can only with difficulty give the settlement solely a 'profane' classification. Ayn-Asil, however, is more clearly domestic, yet has flint knives, although as Midant-Reynes (1998:44) states, there is a far smaller percentage of knives at this site than others. Miller (1985) believes that stone knives found at Karnak were used for temple butchery. However, such context-derived conclusions alone are imprecise. Temples were engaged in what we would classify as secular activity, so these contexts do not prove ritual use.

Until the New Kingdom, flint knives feature in cattle butchery scenes decorating tomb walls. Since faunal remains suggest that little beef was actually consumed, their frequent representation suggests a more than utilitarian purpose. Even without text, one may put two and two together and conclude that knives were employed in ritual butchery.

But were the tools depicted of metal, not stone? Some Egyptologists deny the use of flint knives after the Old Kingdom, others extend use into the Middle Kingdom. In the 5th-Dynasty mastaba of Hetepka 'sparks are seen falling from sharpened knives, indicating that they are of metal' (Martin 1991:12). Martin makes no admission that 'sparks' may be flint spalls. Eggebrecht (1973) posits flint knife use until ousting by metal in the Middle Kingdom. He explains depictions of flint knives on tomb and temple walls from the New Kingdom as archaisms and recognises only one depiction of a New Kingdom flint knife, that from the grave of Tetiky at the beginning of the 18th Dynasty, ca 1550 B.C.E. (Davies 1925: Pl. III). The knife is depicted with a roughened texture resembling flint, rather than the smooth surface of metal (Figure 2.1).

Close investigation of tool sharpening scenes suggests that after the Middle Kingdom the flint knife was no longer the normal butchery implement. For example, the 6th-Dynasty (2345–2181 B.C.E.) tomb of princess Idut at Saqqara shows the striking tool held in the *right* hand and the knife in the *left* hand, as one would expect for a resharpening stone (illustrated in Houlihan 1996:Fig. 14), and similar scenes appear in many other Old Kingdom tombs (for example, Figure 2.2). Such scenes contrast with the New Kingdom knife sharpening scene from the tomb of Khaemwese, where the whetstone is held in the right hand (illustrated in Martin 1991:Fig. 124) and the 25th-Dynasty scenes from the tomb

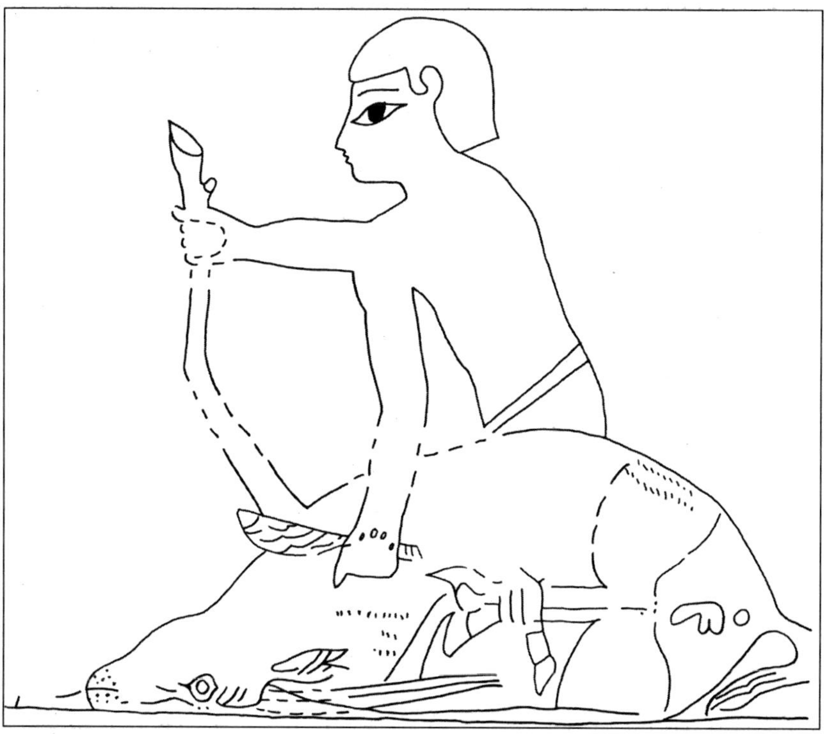

Figure 2.1 • Butchery scene from the Tomb of Tetiky, Thebes, 18th Dynasty (redrawn from photograph in Carnarvon and Carter 1912, Pl. VI.2).

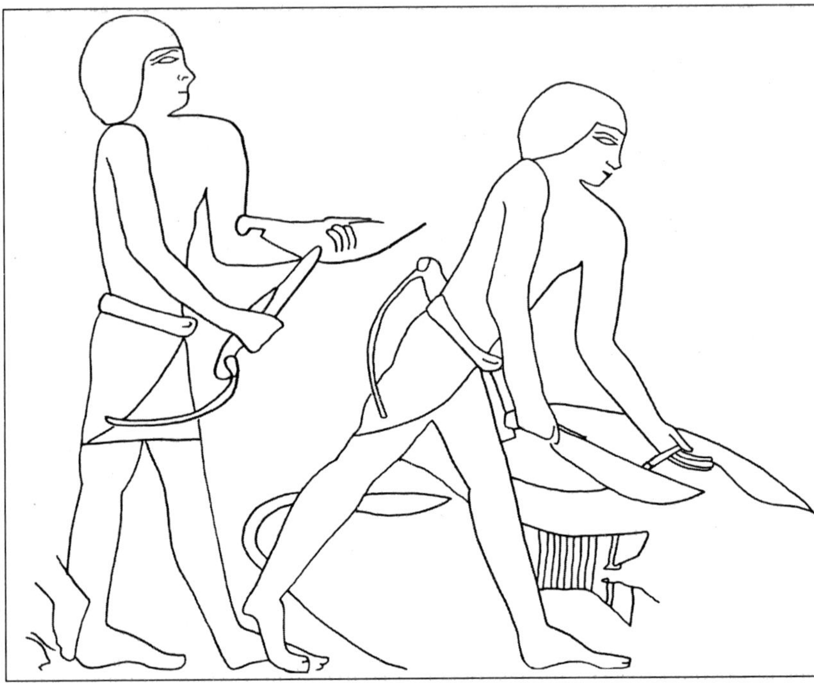

Figure 2.2 • Old Kingdom butchery scene from the Tomb of Ptah-hotep and Ankh-hotep showing the sharpening of a knife (after Montet 1910:44) (reproduced with permission of the IFAO).

Figure 2.3 • Detail of a butchering scene in the 25th-Dynasty Tomb of Mentuemhat (Saint Louis Art Museum); note that the sharpening stone is held with the left hand.

of Mentuemhat at Thebes (Figure 2.3). However, since flint knives are known from New Kingdom contexts, one cannot be sure that they were *never* used in ritual butchery.

What of textual evidence? Flint knives are not explicitly associated with ritual butchery of animals, although they appear in the *Pyramid Texts* as aggressive entities. Text has also been used to identify an archaeological structure associated with knife fragments at the 5th-Dynasty (2494–2345 B.C.E.) 'Shrine of the Knife' of Raneferet (Vachala and Svoboda 1989).

After the New Kingdom, written evidence increasingly suggests that flint knives in particular were considered suitable weapons against malevolent deities such as Apophis. For example, slaughter takes place at midday on a sandbank or at 'Two Knives Lake'. The act is usually taken to be a manifestation of law and order (Assmann 1995:53). Iconography supports this with mythic scenes of various knife-wielding deities.

Text thus complements archaeology, adds specifics, and makes explicit the mythic dimension. What is noticeable is that archaeological evidence of flinty ritual slaughter is particularly manifest in the Old Kingdom, and iconographic evidence continues until the Middle Kingdom. However, textual evidence concerning not ritual slaughter of animals but mythic slaughter of deities grows through the New Kingdom. Perhaps archaeology is more attuned to lived reality? I will return to this point.

Protective Qualities

A miniature Early-Dynastic flint knife found by Petrie in the town of Abydos (Petrie 1902:24, Pl. LI) portrays amuletic, therefore protective, qualities. Although other miniature knives, mainly metal, are known, their funerary contexts may imply afterlife provisioning. Early-Dynastic amuletic flint *psS-kf* are also known (for example, British Museum 37279; Petrie 1902:24 and Pl. LI, 22; Spencer 1980:101 [755], Pl. 79). *PsS-kf* (literally, 'split flint') were originally a type of Pre-Early-Dynastic knife. Later amuletic *psS-kf* are known in other materials (Roth 1992). An interesting ivory object apparently imitating a flint blade inscribed with the name of the 1st-Dynasty king Aha is known from Abydos (Petrie 1901:21, Pl. XXXVI). Amuletic flints of nonknife form occur at Deir el-Medina, presumably New Kingdom in date (Bruyère 1933:7), and a pendant of 'roughly ground and pierced' black flint was found at Tanis, dating to the Ptolemaic Period (Petrie 1885:34). In these two cases at least, we can assume that the material, and not the form, was the protective element. There is a problem, however, in recognising flint as an amuletic material in the published literature, for when it is particularly attractive and thus selected for wear and/or polished it tends to be categorised as 'chalcedony'.

What of iconography? In Old Kingdom funerary contexts, snake hieroglyphs are severed by knife icons, perhaps to protect the dead (Lacau 1914; Ritner 1993:157; Yoyotte 1963:103). Protective knife-wielding deities adorn Middle Kingdom 'apotropaic wands', New Kingdom *Book of the Dead*, 22nd-Dynasty coffins (945–715 B.C.E.), and so on. What would we make of these without text? Given the archaeological occurrence of flint knives at this time, we may envisage that at least some iconographic examples wielded by apotropaic warrior gods were of stone.

Textually, it seems that flint and its form in the knife were both imbued with the essence of protection. A stela of Ramesses II (New Kingdom) reads:

> Sovereign of the Two Lands like Atum
> Wall of flint around the Nile-land [. . .] (Kitchen 1975:87)

Flint knives are used against snake bites (Graves-Brown 2005:61). Other non-snake-related positive values are attested for flint. In the *Book of the Dead*, a staff of flint gives 'the breath of life' (Naville 1886:II, 324). Furthermore, flint appears in the Ptolemaic Edfu mineral lists as an amuletic material (Edfu VI 1 4–5, cited in Aufrère 1983:footnote 47).

Possible Creational Aspects of Flint Knives

Given that flint knives occur frequently in Old Kingdom tombs, both as artefacts and depicted in butchery scenes, and that tombs are monuments replete with rebirth imagery, a creational aspect may be envisaged for the knives. What is the supporting archaeological evidence? Flint was perhaps used in transitional phases of human life-death (Graves-Bown 2006a/b): Roth (1992) suggests the *psS-kf* was employed in cutting the umbilical cord; flint fragments were found inside pots containing the bodies of neonates at Deir el-Medina (Bruyére 1937:12); and flint may have been used in circumcision and embalming. There are few published excavated instances of flint embalming tools such as those from Deir el-Medina (Bruyére 1937:80). Flint embalming instruments from Memphis held by the Berlin Museum were unfortunately lost in the Second World War (Janot 2000:162–63). As to textual evidence, we have one Biblical instance of circumcision using flint (Exodus 4:25) and Herodotus's description of embalmers employing stone knives (Herodotus, Book II, 37; trans. Sélincourt and revised Marincola 1996:143).

Creational aspects of flint may also be evident in two stelae from Deir el-Medina that contain flint nodules. One, now in Cairo (Bruyére 1939:199–200 and Pl. XXIII; Keimer 1940:16–17, Pl. X) contains a flint that may represent the sun emerging on the horizon or the moon within its crescent. The other stela (Cerný 1958: Banks Stela number 4), now at Kingston Lacy (Figure 2.4), also originally had a flint nodule, which I believe represented the sun (Graves-Brown 2006b). Another 19th-Dynasty Deir el-Medina stela in the museum at Turin may possibly be of the same type, although the picture and the description are not decisive (Tosi and Roccatiti 1972:80, no. 50047).

Figure 2.4 • Banks Stela 4, now at Kingston Lacy. The circular feature in the top register, just underneath the winged scarab, is the place where the flint nodule used to be (drawing by Paul Graves-Brown).

Archaeologically, the protuberant nature of the nodules in the two stelae is unusual, perhaps recalling the scarab of New Kingdom and later pylon-shaped pectorals. Studies of these suggest that the pectoral is a functional means of supporting the scarab, though text indicates that the pylon is also replete with symbolism of birth and emergence (Finnestad 1998). The emergent character of the scarab reflects iconographically the sun in the stage of becoming. One is tempted to see a similarity in the flint emerging from the stone.

In ancient Egyptian, the scarab is *khepri*, 'who comes into being', and a creature who, like the sun, travels East to West. These particular nodules come from Western Thebes, a place of rebirth. It might have been possible to conjecture this from archaeology, since the West bank is the site of mortuary temples and tombs of kings and officials from the Middle Kingdom to the end of the Pharaonic Period.

Creational aspects may also be suggested by the pregnant women figurines made by painting natural nodules from Deir el-Medina (Bruyére 1939:144 and Pl. XLV; Keimer 1940) (Figure 2.5). Other similar shaped but undecorated nodules come from Elephantine (Dreyer 1986:96–97, 153, Pl. 57), Timna (Rothenberg 1988:268, Pl. 155) and Mirgissa (Vercoutter 1970:329, Pl. 27. 28). These are not, however, flint, which may suggest that it is the shape, not the material, which was important.

Figure 2.5 • Naturally shaped flint nodule decorated to represent a pregnant woman, from Deir el-Medina (drawing by Paul Graves-Brown).

The only two scenes of flint knife manufacture known from tomb walls both occur in the Middle Kingdom cemetery of Beni Hasan (Griffith 1896). Although tomb scenes tend to be regarded as imagery of rebirth, that only these examples depict lithic technology might suggest that the theme was absent from the standard creational repertoire.

Iconographic and literary sources describe the *psS-kf* used in the 'opening of the mouth' ritual designed to bring statues and human dead to life. However, many other instruments of differing materials were also used (Otto 1960). Using textual evidence, Eyre (2002:83) has pointed out that butchery is used in the *Pyramid Texts* as a metaphor for transformation; therefore, indirectly, the flint knife is implicated in this process.

Non-Utilitarian Forms

Ritual function is often implied by non-utilitarian forms. Amuletic knives have already been discussed. Particularly fragile (unusable) examples (Holmes 1989:282; Kelterborn 1984) and oversize monstrosities, such as that from Hierakonpolis (Adams 1995) and the 72-cm example from Abydos (Hikade 1997), are also known from the Early Dynastic Period (Figure 2.6). There is no direct textual evidence clearly explaining non-utilitarian knives. The 22nd nome name incorporates a knife, and one could imagine large knives on nome standards. Monstrous examples may have featured in temple rituals. Possibly they were considered protective or perhaps associated with ritual sacrifice (discussed below). We must admit, however, that we are on speculative grounds.

Heirlooms

The oversize Early Dynastic knife from Abydos was handed down through generations before final deposition (Hikade 1997). A 5th-Dynasty temple deposit at Elephantine (Dreyer 1986:Fig. 45.351) contains, on

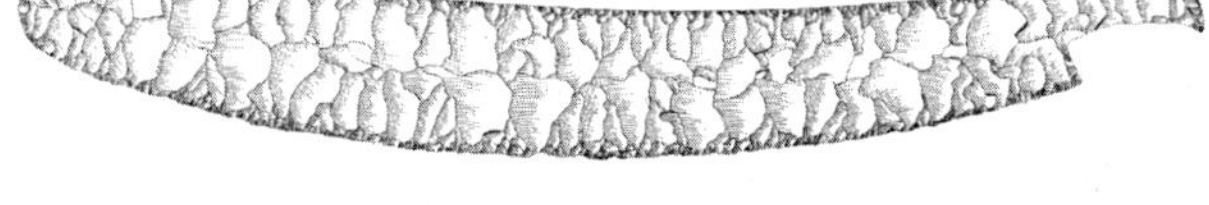

Figure 2.6 • 72-centimetre long Early Dynastic Knife (after Hikade 1997, with kind permission of the author).

typological comparison, at the latest a 3rd-Dynasty knife. Thus the knife was kept for at least 100 years before being deposited. Given the fragility of such specimens, careful curation must suggest a more than functional artefact. Possible reasons are expanded in text – the Egyptians imbued the distant past with special qualities, as 'the time of the gods', a perfect time, and stone represented this time (Ogdon 1990:17–22; Wilkinson 1999:89). A temple inventory of Abu Sir (ca 2427 B.C.E.) lists a number of *psS-kf* knives, mostly broken, some repaired (Posener-Kriéger and de Cenival 1968:115, Fig. 20). It seems surprising that rich temples would be unable to acquire new ritual items, so use of damaged pieces may imply respect for the past.

Text-Centred Evidence

Textual exploration of the symbolism of lithics has been admirably undertaken by Aufrère (1983, 1991, 2001, and so on). He demonstrates ideological connotations surrounding flint: its association with the Eye of Re; with fire; with certain archer goddesses; the Eastern Desert; meteoric iron; perhaps with the god of chaos, Seth; with ritual killings of Apophis, the enemy of order; and so on. These aspects are now summarised (discussed more thoroughly in Graves-Brown 2005), first in connection with text and then with iconography and archaeology.

Flint from Re to Kill His Enemies

Textual evidence reveals that flint, like other stones, was linked with the sun god Re, more indirectly with the Eye of Re. The Eye is affiliated with kings prospecting for minerals in foreign parts and so forth (Galán 1994:74). Lion-headed goddesses personifying the Eye of Re star in stories where they, too, are brought back from the Eastern Desert. Serpents represent the solar eye, usually as a protector and are conceptually linked with flint. Flint is part of the metaphoric set *flint-fire-Re-lioness-goddess.*

What can archaeology add? Regarding the Re pedigree, very little. We may be on slightly firmer ground regarding killing of enemies. Although we cannot know for certain, the label of Aha from Abydos and that of Djer from Saqqara possibly show human sacrifice using knives (Wilkinson 1999:266, Fig. 8.2). Both date to the Early Dynastic Period. The only other evidence for human ritual killings – or as Ritner argues, an execration rite – using flint has been posited by Vila for Mirgissa (Ritner 1993:163, 170–71; Vila 1973).

Fire

The archer goddess Sekhmet is very often associated with fire. Here, as elsewhere, flint and fire are used in tandem to destroy wrongdoers.

In iconography, solarised deities such as Bes carry knives. Magical serpents, female warriors (often taking a lioness form), doorkeepers of the otherworld are all interconnected with knives and fire. Metaphors surrounding flint knives and fire are similar. Although the knives are not always clearly flint, they are not infrequently so.

The word used in connection with flint knife manufacture or sharpening at Beni Hasan is *sxt* ('to strike/make'). In the New Kingdom, this word also occurs in connection with fire and seems to mean 'to light a fire'. The *Pyramid Texts* (ca 2350 B.C.E.) describe a knife (presumably flint) that is said to 'lick' *nsb* (Pyr. 228), a word that is also used for fire. Fire and flint are both 'sharp'. The fire or knife determinative can both be used for the word 'to brand', *Abw*.

In text, *ds* and flame together smite the enemies of Re, for example, in the 4th-century B.C.E. *Bremner-Rhind Papyrus.* Additionally, Horus of Letopolis has two eyes of flint or two eyes of flame in Chapter 125 of *Book of the Dead.*

Explanation for fire affiliations seems obvious – flint emits sparks when struck with iron. This might suggest strike-a-lights, especially given additional textual connection with meteoric iron. Strike-a-lights are not known archaeologically in Egypt – surprising, perhaps, since manufacture of flint tools must have occasionally given rise to sparks and because of the belief that there was fire in flint. However, Runnels (1994) claims that although it is relatively simple to make fire with especially prepared iron strikers, it is not so easy with natural minerals, as would have been employed until ca 700 B.C.E. in Egypt. She suggests that the fire drill is an earlier method of making fire, and examples of such drills are known in Egyptian iconography and archaeology.

The suggestion that butchery is part of the transformation process was pointed out above. The Lake of Fire in the afterworld is also a means of transformation (Altenmüller 1966; Bourghouts 1971:103–04). Interestingly, in some texts from the 19th Dynasty, the Lake of (Two) Knives can be substituted with the Lake of Fire, although the knives (*dsds*) are not necessarily of flint, but could be metal. In hermopolitan myth, the sun god is born from the Lake of Knives (Altenmüller 1966).

Snakes

In the Middle Kingdom *Coffin Texts* (160, 375b), a snake whose front part is made of flint is described on the mountain of *BAXw*. The snake is called 'who is on his mountain, who is in his flame'. The snake and flame connection appears obvious. *BAXw*, a mythical place in the East, was a realm of danger and difficulty as well as rebirth (Posener 1965:76).

According to the literature, snakebites are treated with flint knives, and Apophis, who takes the form of a snake, can be killed with a flint knife.

In the *Pyramid Texts* (Utterance 228), a flint knife is described which swallows, licks, and is particoloured, all snake imagery (Ritner 1993:97).

Can archaeology augment text here? In the Late Predynastic, knife handles were adorned with elephant and serpent, or plain serpent, motifs (Johnson 1990:38–42, 44–45), although an ivory comb and a macehead were also so decorated. A 1st-Dynasty flint snake is known from Coptos (Petrie 1914:25, Pl. XII 96b). Near the Predynastic settlement of Mahasna, a deposit of natural flints was unearthed, some of which were snakelike in appearance (Garstang 1903:7). However, these would appear insignificant without textual data, since non-ophidian flint animals are known.

Desert

Given the occurrence of desert flint mines, the desert imagery is unsurprising. The Eastern desert is particularly poignant, a birthplace of the sun and gods, although flint mines also occur in the West.

Seth is a desert god, though usually associated with the Western Desert. In the *Pyramid Texts* (Utterance 665 [1906], Utterance 674 [1999]), and the *Coffin Texts* (1491), *ds* is said to come from Seth.

From literature we deduce that the lion is a solar animal, therefore connected with quartz in particular and perhaps with stone more generally. Lions are associated with the desert, and magical lions have claws of flint (Alliot 1949–1954:II, 752–53; Chassinat 1931:VI, 127, 11). The Middle Kingdom ivory 'magic wands' often depict lion-headed gods carrying knives. Fiery lioness goddesses connected with the wandering Eye, brought back from the Eastern Desert, have already been mentioned.

Posener (1965) explains, using iconography and text, that the left/East is both a place of birth from where the sun rises and a place of danger. In opposition, the right/West is the place of honour. The Eastern Desert from which flint ideally originates is associated with difficulty and danger as well as rebirth. Almost invariably in butchery and reaping scenes, humans (men) are shown holding knives with their right hands (except when sharpening them). Deities are ambidextrous, and the Re cat often uses his left paw. This may be no more significant than an illustration of the special skills of the gods.

BiA

In literature, flint and *biA* (usually translated as 'meteoric iron') are frequently paralleled. For example, in the *Abu Sir Papyrus*, flint (*kAf*) appears as a word between *biA* and quartz. Amulets of *ds* and *biA* are mentioned at Ptolemaic Edfou (Chassinat 1931:VI, 299, 12). Wainwright (1932), using text largely of the New Kingdom and later, suggested that *ds* and *biA* were both connected with storms, with Letopolis, and overthrowing Apep.

Roth (1993:72) theorises that stones of odd shapes were thought to originate in heaven. These include flints, fossils, and *biA*, which also means 'marvel, miracle'. Does archaeology support this? Both flint and *biA* are collected for their odd shapes; both used for manufacture of amulets. Both the flint *psS-kf* and the iron adze are used in the 'opening of the mouth' ritual. However, one has to accept that, without text and iconography, we could not know the ceremony, that the adze is not always of meteoric iron, and that later *psS-kf* are not always of flint. Roth (1992:137) also describes 5th- and 11th-Dynasty amuletic *psS-kf* with a blade of meteoric iron, although she rightly states that the iron may have been chosen for its rarity value.

Discussion: Integrating Text and Archaeology Using Metaphor Theory

With few exceptions, 20th-century scholars seem largely to have followed a Cartesian dualist approach to text and archaeology. Wrapped in 'scientific ideals', scholars saw texts as of the mind, as precise, as useful vehicles through which to convey the 'higher' expressions of human creativity, such as religion. Archaeology was an untidy product of lived-in reality expressing the 'lower' orders of ecology, production, and technology. Although there is some truth in this, as usual reality is a little more muddied.

From the preceding material, one can see that texts tend to give more specific information – for example, linking specific deities with flint. Archaeological materials seem particularly useful in societies where there is a non-existent or low level of literacy, and this includes Egypt, where the silent majority can be studied only through archaeology. Together, text and archaeology show that flint knives have many ideological meanings; they are multivocal.

In this section, I tentatively propose that differences between textual and archaeological evidence may correlate with those between explicit (live) and implicit (dead) metaphor, as postulated by metaphor theory. Instead of simply uncovering evidence in one area and then digging for confirmation in another, it might be profitable to consider the difference and dialectic between the two, and perhaps elucidate ideological evolution. I would like to explore the possibility of ideological development of the metaphor from its early explicit and then implicit inception in the object, to its explicit manipulation in text.

Very briefly, metaphor theory sees thought as embodied and imaginative, rather than literary. Concepts not directly grounded in experience are explained through metaphors. This includes not only the ornamental but also the commonplace and scientific. Use of the term 'metaphor' is here not to be seen in a purely linguistic light following Chomsky and the structuralists. Rather, it derives from the word as employed by those such as the art historian Gombrich (1972:165–91) and further developed by the linguist Lakoff (1987) and the philosopher Johnson (1987; see also Lakoff and Johnson 1999). The idea has recently been used archaeologically by Tilley (1999). Since, strictly speaking, this application of metaphor also includes metonym and synecdoche, 'metaphor' should possibly be replaced by the term 'trope'. However, the expression 'metaphor theory' has already entered popular usage.

Unconscious (implicit) use of metaphor is sometimes described as 'frozen' or 'dead'. In literature, this use tends to refer to tropes that move from explicitly manipulated to implicitly accepted forms, but in more general terms they can also move from implicit to explicit. Most studies of metaphor focus on short-term effects, the metaphor moving from explicit to implicit. Archaeology (together with history) may have the advantage in allowing study of metaphor over a long period, and thus it may be possible to see 'birth', 'death', and 'rebirth' of a metaphor, moving from explicit to implicit and back to explicit. For example, we may postulate that the use of flint knives as a symbol of creation and protection arose in the Pre- and Early Dynastic. The oversize and fragile forms may suggest a deliberate, that is explicit, manipulation of metaphors. By the Old Kingdom, butchery scenes show implicit metaphors. The knife is present, there is indication of its importance, but it is more subtly expressed. When flint knives decrease in everyday use, owing to increase in metal, the flint knife – we see from text – continues or perhaps even strengthens as a metaphor for protection. The difference between the early and the later explicit use of flint is that the former involves objects and, the latter, text. I suggest below that this later textual use could be viewed as explicit, especially where text concerns the mythic.

We can look at this development in more detail. It is frequently stated that the medium harbouring the strongest cultural metaphor is the primary medium (Boas 1955; Roe 1995). It is difficult to identify the primary medium owing to the bias of preservation and selection, but the ubiquity of the material in Egypt suggests that in Predynastic Egypt flint could have fulfilled that role.

State formation must have engendered cultural upheaval. As stated by Tilley (1999:6), most scholars agree that literal metaphors are created because of relational problems between the world of words and the world of things. Metaphors are used to try to capture the otherwise unexplainable. One would expect important new metaphors to occur at a time of change. Indeed, state formation coincides with the manufacture of beautifully elaborate and oversize flint knives, perhaps demonstrating, in a physical sense, ideals of the early state. Their size and fragility suggest a symbolic or metaphoric purpose. The extreme nature of these tools makes it difficult to believe that their symbolism can have been other than explicit or deliberate. These categories were perhaps manipulated by the 'intellectual elite', as Kus (1992) suggests for state societies.

It could be argued that ritual, by its very nature, is explicit; therefore, Early Dynastic evidence for ritual flint indicates explicit use of metaphor. However, ritual may sometimes be ethically understood only as explicit (Gell 1975:211). For medieval monks, performance of rites was not for their symbolic meaning; the practice itself was essential (Asad 1993).

The earliest textual references to flint knives occur in the *Pyramid Texts*, first written down in the late Old Kingdom, although usually believed to have been composed earlier, ca 3000 B.C.E. (Mercer 1952:1) – that is, around the time of the early elaborate flint knives. There is thus a concentration of literary and archaeological evidence for the symbolic use of the knife in this period.

Are there differences in the way material and literal metaphors operate? Material metaphors work first through internal qualities such as shape, form, and texture and second in actional and biographical context (Tilley 1999:263–64). Also, because they are solid, they can be read immediately, but their meaning is more subtle. With solid metaphors, mixed metaphors are acceptable. Finally, 'solid metaphors are not substitutes for linguistic metaphors or translations of them into material form. They act most subtly and powerfully precisely when they are not linguistically translated, at a non-discursive level of consciousness and as part of the routinization of action' (Tilley 1999:263). Difference, however, does not equate with inferiority.

Some might say that literary and concrete metaphors are *very* different. Brunner-Traut stated that in the construction of the state the Egyptians established a new mental order and 'released themselves from their attachment to the object' (Brunner-Traut 1986:439). This new mental order at the birth of state is explored by Goldwasser (1995). She suggests that advanced metaphor first appears on the Narmer Palette and the Libyan Palette (Early Dynastic). I agree that the introduction of a written language may allow a perceived separation between abstract and more concrete thought, and perhaps ideas in literate societies may be termed more metaphoric. However, true metaphoric thinking does exist in pre-literate societies (Tilley 1999:68–72). Following Finnegan (1973), I do not believe in the great divide between literate and nonliterate people, but rather this is a difference in degree between the explicit and implicit metaphor (*sensu* Goldwasser 1995). The absolute dualism between the world of objects and that of writing was not recognised by the Egyptians themselves, who clearly believed that the writing of words could make them reality (hence the mutilation of 'dangerous' symbols in Old Kingdom tombs). Indeed Egyptologists such as Alan Gardiner (1961) saw the Egyptian language as being particularly 'concrete'. On the one hand, we have highly symbolic objects and, on the other hand, a 'concrete' language. I see solid and literal theology as a continuum, although with literature leaning toward the explicit.

The Old Kingdom sees a decline in quality of flint knives, for which various alternative explanations may be postulated. It could represent a move from labour-intensive to mass-produced forms or cultural collapse. Maybe symbolic aspects of flint were explored in other materials such as metal. Gero (1989:104) has suggested that additive technologies such as pottery might be better suited to conveying social messages than subtractive technologies such as lithics.

The Old Kingdom does not see explicit symbolic use of the material form of the flint knife. However, the use of the flint knife in ritual butchery, as shown in iconography, does suggest belief in its creational aspects. This subtlety suggests an implicit use of metaphor. At this stage iconography does not include depiction of explicit deities. Old Kingdom textual references to flint knives largely come from the *Pyramid Texts*, though these were probably composed in the Early Dynastic.

During the Middle Kingdom, flint is rarely depicted in iconography and occurs less on funerary sites and more on domestic sites. Its importance in the literature continues, being incorporated into other symbolic constellations surrounding human procreation, the body and so forth. Its durable quality, sharpness, protective symbolism, and solar connections mean that in turn it was used metaphorically outside its manufacturing or immediate use categories.

From the New Kingdom, there is little archaeological and iconographic evidence for 'ritual' use of flint; the focus of evidence centres on text where knives appear used against malevolent deities. Flint knives are used, however, in other possible 'religious' practices – for example, against snake bites as attested by the *Brooklyn papyrus* (30th Dynasty or Early Ptolemaic) and for various ailments in the *Ebers papyrus* (ca 1530 B.C.E.) and *Papyrus Ramesseum* V (ca 1795–1550 B.C.E.). We would not expect archaeological evidence for such ephemeral use. What may we infer? Naturally, more literature survives from later dates, but then why not archaeology and iconography? Perhaps text is simply archaic, and possibly other media expressed ideology. It is possible that, as Assmann (2001) suggested, religion became secularised, or maybe we are seeing the death of a metaphor? The suggestions are not mutually exclusive.

The argument that we are witnessing a move from a more implicit metaphor of the object toward the explicit metaphor of the written word may be advanced through examination of the impact of writing upon thought. A number of authors have discussed the effects of text upon society, including Goody, Ong, and McLuhan. This is not the place to discuss their work in detail, but the setting down of activity into words does to some extent remove an act from its embodied nature; it becomes in some sense more numinous, mythic, an element of the explicit metaphor. Therefore, while acknowledging that language must of necessity be common to both, we see a potential difference in meaning between the act as embodied and the act as written. Although archaeology can deal with 'taken for granted' aspects of daily life (yet it may include elements of the sacred), text records the deliberate thought processes. This does not mean there were no discursive thoughts without text, but simply that they were not recorded.

Text seems to concern mythic rather than ritual slaughter, not of real animals but deities. Cassirer (1944) and Avis (1999:117) state that myths are ever myths only once they cease to become credible. Perhaps, by the New Kingdom, flinty metaphors, already implicit, had become frozen as myth. Flint, as a material representing creation and protection, was now a metaphor that was no longer obvious, as flint itself was not so commonplace. It is not abandoned, however, but rather enters the realm of myth. There may be various reasons why it is not abandoned, one being that, as stated above, stone may represent the archaic or perfect time of the gods.

I believe that the move from implicit to explicit metaphor as shown by use of flint in ideology is also linked to a move from profane to sacred. When metaphor is grounded in the unconscious it remains implicit; to outside observers it appears concrete, even profane. When it is explicit, deliberate, conscious, we can explain it as ideology, using words such as 'abstraction' or 'sacred'. At first, flint had been shown to be embedded and part of the belief system of the ancients. Later, it appears more in text and less in archaeology, in objects. Metaphor or myth may or may not be deliberately manufactured and may or may not be reflexively understood.

In sum, an examination through text and archaeology of the ideology surrounding flint in Dynastic Egypt shows that, at different periods, flint is used symbolically in a variety of ways, implicitly and explicitly. An exploration of the long-term changes in metaphoric use of flint suggests that the difference between text and archaeology roughly relates to the difference between the ideal (numinous) explicit metaphor, and the embodied implicit metaphor.

References

Adams, B. 1995. *Ancient Nekhen: Garstang in the City of Hierakonpolis.* New Malden: SIA.

Alliot, M. 1949–1954. *Le culte d'Horus à Edfou au temps des Ptolémées.* Cairo: L'Institut Français d'Archéologie Orientale.

Altenmüller, H. 1966. *Zeitschrift für Ägyptische Sprache und Altertumskunde* 92: 86–95.

Asad, T. 1993. *Genealogies of Religion: Discipline and Reasons of Power in Christianity and Islam.* Baltimore: John Hopkins University Press.

Assmann, J. 1995. *Egyptian Solar Religion in the New Kingdom: Re, Amun and the Crisis of Polytheism.* London: Kegan Paul International.

———. 2001. *The Search for God in Ancient Egypt* (translated from the German by D. Lorton). Ithaca, NY: Cornell University Press.

Aufrère, S. 1983. 'Caractères principaux et origine divine des minéraux', *Revue d' Égyptologie* 34:3–21.

———. 1991. *L'univers minéral dans la pensée Égyptienne* (2 vols.). Cairo: L'Institut Français d'Archéologie Orientale.

———. 2001. 'The Egyptian temple, substitute for the mineral universe', in W. V. Davies (ed.), *Colour and Painting in Ancient Egypt.* London: British Museum Press, 158–63.

Avis, P. 1999. *God and the Creative Imagination: Metaphor, Symbol and Myth in Religion and Theology.* London: Routledge.

Baines, J. 1983. 'Literacy and ancient Egyptian society', *Man* 18(3):572–99.

Boas, F. 1955. *Primitive Art.* New York: Dover Publications.

Bourghouts, J. F. 1971. 'Excurses II: The sAb.t-snakes', in *The Magical Texts of Papyrus Leiden. Papyrus I.348* (J.F. Bourghoutsl), Leiden: Brill, Rijksmuseum van Oudheden te Leiden, 200–09.

Brunner-Traut, E. 1986 [1919]. 'Aspective', in H. Schafer (ed.) [J. Baines, trans.], *Principles of Egyptian.* Oxford: Griffith Institute, 421–47.

Bruyère, B. 1933. *Rapport sur les foulles de Deir el Médineh (1930).* Cairo: L'Institut Français d'Archéologie Orientale.

———. 1937. *Rapport sur les foulles de Deir el Médineh (1933–1934).* Cairo: L'Institut Français d'Archéologie Orientale.

———. 1939. *Rapport sur les foulles de Deir el Médine (1934–1935).* Cairo: Institut Français d'Archéologie Orientale.

Carnarvon, Earl of, and H. Carter. 1912. *Five Years' Explorations at Thebes: A Record of Work Done 1907–1911.* Oxford: Oxford University Press.

Cassirer, E. 1944, *Essay On Man.* New York: Bantam Books.

Cerný, J. 1958. *Egyptian Stelae in the Bankes Collection.* Oxford: Griffith Institute.

Chassinat, É. 1931. *Le Temple d'Edfou*, 6. Cairo: Institut Français d'Archéologie Orientale.

Davies, N. G. 1925. 'The tomb of Tetaky at Thebes (no. 15)', *Journal of Egyptian Archaeology* 11:10–17.

Deines, H., H. Grapow, and W. Westendorf. 1958. *Grundriss der Medizin der Alten Ägypter: Übersetzung der Medizinischen Texte* (Vol. 4/1). Berlin: Akademie Verlag.

Dobres, M-A. 2000. *Technology and Social Agency.* Oxford: Blackwell.

Dreyer, G. 1986. *Elephantine VIII, Der Tempel der Satet.* Mainz am Rhein: P. von Zabern.

Eggebrecht, A. 1973. *Schlachtungsbräuche im alten Ägypten und ihre Wiedergabe im Flachbild bis zum Ende des Mittleren Reiches.* University of München, Unpublished PhD Thesis.

Eyre, C. 2002. *The Cannibal Hymn: A Cultural and Literary Study*. Liverpool: Liverpool University Press.

Faulkner, R. O. 1969. *The Ancient Egyptian Pyramid Texts Translated into English*. Oxford: Clarendon Press.

Finnegan, R. 1973. 'Literacy versus non-literacy: The Great Divide?', in R. Horton and R. Finnegan (eds.), *Modes of Thought: Essays on Thinking Western and Non-Western Societies*. London: Faber and Faber, 112–45.

Finnestad, R. B. 1998. Temples of the Ptolemaic and Roman Periods: Ancient Traditions in New Contexts, in B. E. Shafer (ed.), *Temples of Ancient Egypt*. New York: I. B. Tauris, 185–236.

Galán, J. M. 1994. 'The Stela of Hor in context', *Studien zur Altägyptischen Kultur* 21:65–79.

Gardiner, A. H. 1961. *Egypt of the Pharaohs*. Oxford: Oxford University Press.

Garstang, J. 1903. *Mahâsna and Bêt Khallâf*. London: Egyptian Research Account.

Gell, A. 1975. *Metamorphosis of the Cassowaries: Umeda Society and Ritual*. London: Athlone.

Gero, J. M. 1989. 'Assessing social information in material objects: How well do lithics measure up?', in R. Torrence (ed.), *Time, Energy and Stone Tools*. Cambridge: Cambridge University Press, 92–105.

Giddy, L. 1999. *Kom Rabi'a: The New Kingdom and Post-New Kingdom Objects*. London: Egypt Exploration Society.

Goldwasser, O. 1995. *From Icon to Metaphor: Studies in the Semiotics of the Hieroglyphs*. Fribourg: Orbis Biblicus et Orientalis 142, University Press.

Gombrich, E. H. 1972. *Symbolic Images: Studies in the Art of the Renaissance*. London: Phaidon.

Goody, J. 1987. *The Interface between the Written and the Oral*. Cambridge: Cambridge University Press.

Graves-Brown, C. A. 2005. 'The Spitting Goddess and the stony eye: Divinity and flint in Pharaonic Egypt', in K. Piquette and S. Love (eds.), *Current Research in Egyptology 2003: Proceedings of the Fourth Annual Symposium. University College London 2003*. Oxford: Oxbow Books.

———. 2006a. 'Flint and human life-crisis in Pharaonic Egypt', in R. J. Dann (ed.), *Current Research in Egyptology 2004: Proceedings of the Fifth Annual Symposium. Durham 2004*. Oxford: Oxbow Books, 68–83.

———. 2006b. 'Emergent flint', in K. Szpakowska (ed.), *Through a Glass Darkly: Magic, Dreams, and Prophecy in Ancient Egypt*. Swansea: The Classical Press of Wales, 47–62.

Griffith, F. L. 1896. *Beni Hasan* (Part III). London: Egypt Exploration Society.

Harris, J. R. 1961. *Lexicographical Studies in Ancient Egyptian Minerals*. Berlin: Akademie Verlag.

Hawkes, C. 1954. 'Archaeological theory and method: Some suggestions from the Old World', *American Anthropologist* 56:155–68.

Hikade, T. 1997. 'Ein aussergewöhnliches Silexmesser aus Abydos', *Mitteilungen des Deutschen Archäologischen Instituts Abteilung Kairo* 53:85–89.

Herodotus. 1996. The Histories (trans. J. Marincola, rev. A. Sélincourt). London: Penguin Classics.

Holmes, D. L. 1989. *The Predynastic Lithic Industries of Upper Egypt: A Comparative Study of the Lithic Traditions of Badari, Nagada and Hierakonpolis* Cambridge Monographs in African Archaeology 33. BAR International Series 469 (2 vols.), Oxford.

Houlihan, P. F. 1996. *The Animal World of the Pharaohs.* London: Thames and Hudson.
Janot, F. 2000. *Les instuments d'embaumement de l'Égypte ancienne.* Cairo: Institut Français d'Archéologie Orientale.
Johnson, M. 1987. *The Body in Mind: The Bodily Basis of Meaning, Imagination and Reason.* Chicago: University of Chicago Press.
Johnson, S. B. 1990. *The Cobra Goddess of Ancient Egypt.* London: Kegan Paul.
Keimer, L. 1940. *Jeux de la nature: Retouchés par la main de l'homme, provenant de deir el-Médineh (Thèbes) et remontant au nouvel-empire.* Cairo: Études d'Égyptologie II. L'Institut Français de Archéologie Orientale.
Kelterborn, P. 1984. 'Towards replicating Egyptian predynastic flint knives', *Journal of Archaeological Science* 11:433–53.
Kitchen, K. 1975. *Ramesside Inscriptions.* Oxford: Blackwell.
———. 1978. *Siedlungsfunde aus dem frühen Alten Reich in Giseh: Österreichische Ausgrabungen 1971–1975.* Wien: Österreichische Akademie der Wissenschaften.
Kromer, K. 1978. *Siedlungsfunde aus dem Frühen Alten Reich in Giseh Österreichische Ausgrabungen 1971–1975.* Vienna: Österreichischen Akademia.
Kus, S. 1992. 'Toward an archaeology of body and soul', in J.-C. Gardin and C. S. Peebles (eds.), *Representations in Archaeology.* Bloomington: Indiana University Press, 168–77.
Lacau, P. 1914. 'Suppressions et modifications de signes dans les textes funéraires', *Zeitschrift für Ägyptische Sprache und Altertumskunde* 51:1–64.
Lakoff, G. 1987. *Women Fire and Dangerous Things.* Chicago: University of Chicago Press.
Lakoff, G., and M. Johnson. 1999. *Philosophy in the Flesh: The Embodied Mind and Its Challenge io Western Thought.* New York: Basic Books.
Lemonnier, P. 1986. 'The study of material culture today: Toward an anthropology of technical systems', *Journal of Anthropological Archaeology* 5:147–86.
Lemonnier, P. (ed.). 1993. *Technological Choices. Transformation in Material Cultures since the Neolithic.* London: Routledge.
McLuhan, M. 1962. *The Gutenberg Galaxy.* Toronto: University of Toronto Press.
Martin, G. T. 1991. *The Hidden Tombs of Memphis.* London: Thames and Hudson.
Mercer, S. A. B. 1952. *The Pyramid Texts in Translation and Commentary,* Vol. I. New York: Longmans Green.
Michael, M. 2000. *Reconnecting Culture Technology and Nature: From Society to Heterogeneity.* London: Routledge.
Midant-Reynes, B. 1981. 'Les noms du Silex en Égyptien', *Revue d' Égyptologie* 33:39–45.
———. 1987. 'Contribution à l'etude de la societé predynastique: Le cas du Couteau "Ripple-Flake"', *Studien zur Altägyptischen Kultur* 14:185–224.
———. 1998. *Le silex de 'Ayn-Asil Oasis de Dakhla-Balat.* Documents de fouilles de l'IFAO 34. Cairo: Institut Français d'Archéologie Orientale.
Miller, R. 1985. 'Lithic technology in East Karnak', *Journal of the Society for the Study of Egyptian Antiquities* 13:228–36.
Montet, M. P. 1910. 'Scènes de boucherie dans les tombes de l'Ancien Empire', *Bulletin de l'Institut Français d'Archeologie Orientale* 7:41–65.
Morschauser, S. 1991. *Threat-Formula in Ancient Egypt.* Baltimore: Halgo Inc.
Naville, E. 1886. *Das Ägyptische Totenbuch der XVIII. bis XX. Dynastie / Aus verschiedenen Urkunden zusammengestellt und herausgegeben von Édouard Naville.* Berlin: Asher.
Needler, W. 1956. 'A flint knife of king Djer', *Journal of Egyptian Archaeology* 42:41–44.

Ogdon, J. R. 1990. 'Some reflections of the meaning of "megalithic" cultural expression in ancient Egypt (with reference to the symbolism of stone)', *Varia Aegyptica* 6:17–22.

Ong, W. J. 1982. *Orality and Literacy.* London: Methuen.

Otto, E. 1960. *Das Ägyptische Mundöffnungsritual* (2 vols.). Wiesbaden: Ägyptologische Abhandlungen 3, Harrassowitz.

Parkinson, R. B. 2002. *Poetry and Culture in Middle Kingdom Egypt: A Dark Side to Perfection.* London: Continuum.

Petrie, W. M. F. 1885. *Tanis.* London: Egypt Exploration Fund and Truber and Co.

———. 1901. *The Royal Tombs of the Earliest Dynasties* (Part II). London: The Egypt Exploration Fund.

———. 1902. *Abydos I.* London: Egypt Exploration Fund.

———. 1914. *Amulets, Illustrated by the Egyptian Collection in University College, London.* London: Constable.

Pfaffenberger, B. 1988. 'Fetishized objects and humanized nature: Towards an anthropology of technology', *Man* 23:236–52.

———. 1992. 'Social anthropology', *Annual Review of Anthropology* 21:491–516.

Posener, G. 1965. 'Sur l'orientation et l'ordre des points cardinaux chez les Égyptiens', in S. Schott (ed.), *Göttinger Vorträge vom Ägyptologischen Kolloquium der Akademie am 25. und 26. August 1964.* Nachrichten der Akademie der Wissenschaften in Göttingen. Philologische-Historische Klasse, Göttingen:69–78.

Posener-Kriéger, P., and J. L. de Cenival. 1968. *Hieratic Papyri in the British Museum, 5th Series: The Abu Sir Papyri.* London: British Museum.

Roe, P. G. 1995. 'Style, society, myth and structure', in C. Carr and J. E. Neitzel (eds.), *Style, Society, and Person: Archaeological and Ethnological Perspectives.* New York: Plenum Press, 27–76.

Ritner, R. K. 1993. 'The mechanics of ancient Egyptian magical practice', *Studies in Ancient Oriental Civilization No. 54.* Chicago: The Oriental Institute of the University of Chicago.

Roth, A. M. 1992. 'The *PsS-kf* and the "Opening of the Mouth" Ceremony: A ritual of birth and rebirth', *Journal of Egyptian Archaeology* 78:113–47.

———. 1993. 'Fingers, stars, and the "Opening of the Mouth": The nature and function of the ntrwj-blades', *Journal of Egyptian Archaeology* 79:57–79.

Rothenberg, B. 1988. *The Egyptian Mining Temple at Timna* (Vol. I). London: Institute for Archaeo-Metallurgical Studies Institute of Archaeology, University College London.

Runnels, C. 1994. 'Tinderflints and firemaking in the historical period', *Lithic Technology* 19(1):7–16.

Saad, Z. 1951. 'Some of the tombs explored and objects found during the fourth season (1945–1946)', in *The Royal Excavation at Helwan*, Suppl. ASAE 14:1–44.

Schmidt, K. 1992. 'Tel Ibrahim Awad: Preliminary report on the lithic industries', in E. C. M. Brink (ed.), *The Nile Delta in Transition: 4th–3rd Millennium BC.* Tel Aviv: Brink, 79–96.

Shirai, N. 2005. 'Bifacial technology, socioeconomic competition, and early farming and herding in the Fayum', in K. Piquette and S. Love (eds.), *Current Research in Egyptology 2003: Proceedings of the Fourth Annual Symposium University College London 2003.* Oxford: Oxbow Books, 135–48.

Spencer, A. J. 1980. *Catalogue of Egyptian Antiquities in the British Museum (Vol. V): Early Dynastic Objects.* London: British Museum.

Spurrell, F. C. J. 1891. 'The stone implements', in W. H. F. Petrie, *Illahun, Kahun and Gurob.* London: Egypt Exploration Society, 51–56.

Tilley, C. 1999. *Metaphor and Material Culture*. Oxford: Blackwell.

Tillmann, A. 1992. *Die Steinartefakte des dynastischen Ägypten, dargestellt am Beispiel der Inventare aus Tell el-Dab'a und Qantir*. Unpublished PhD Thesis, University of Tübingen.

Tosi, M., and A. Roccatiti. 1972. *Stele e altre epigrafi di Deir el Medina. (n. 50001–50262): Pubblicate con il contributo del Consiglio nazionale delle ricerche*. Torino: Edizioni d'arte Fratelli Pozzo.

Vachala, B., and J. Svoboda. 1989. 'Die Steinmesser aus Abusir: Ein Artefakt aus der Sicht von zwei historischen Disziplinen', *Zeitschrift Altägyptischer Sprache* 116:174–81.

Vila, A. 1973. 'Un rituel d'envoûtement au Moyen Empire Égyptien', in M. Sauter (ed.), *L'Homme, hier et aujord'hui: Recueil d'études en hommage à André Leroi-Gourhan*. Paris: Éditions Cujas, 625–39.

Vercoutter, J. 1970. *Mirgassa* (Vol. 1). Paris: Mission Archéologique Française au Soudan.

Wainwright, G. A. 1932. 'Letopolis', *Journal of Egyptian Archaeology* 18:159–72.

Walker, W. H. 2001. 'Ritual technology in an extranatural world', in M. B. Schiffer (ed.), *Anthropological Perspectives on Technology*. Dragoon, AZ: Amerind Foundation, 87–107.

Wilkinson, T. A. H. 1999. *Early Dynastic Egypt*. London: Routledge.

Yoyotte, J. 1963. 'Etudes géographiques II', *Revue d' Égyptologie* 15:187–19.

CHAPTER 3

Cuneiform Glass Texts: A Question of Meaning

ANDREW J. SHORTLAND

ABSTRACT *Uncovered in the mid-19th century by Sir Henry Layard and his assistants, the library of king Assurbanipal (668–627 B.C.E.) in Nineveh contained a collection of some 25,000 fragments of clay tablets, the largest accumulation of such documents known. Among these documents were a number related to science, mathematics, medicine, and the procedures for the production of glass, perfumes, and other materials. Using the glass texts as an example, this chapter places these 'procedural texts' in the cultural and literary traditions of the period. It goes on to question their utility: are they originals or copies? recipes or dictionaries? ritual or practical? or are they some combination of all of these, their function changing through the life-history of the text? Finally, an assessment is made as to how much, if anything, these texts can tell us about the material, methodological, and ritual practices of glassmakers in the 2nd millennium B.C.E.*

Introduction

Over the last 20 years or so of research into the production of ancient glass, the cuneiform glassmaking texts have largely been ignored. The seminal work on the subject, *Glass and Glassmaking in Ancient Mesopotamia*, was completed in 1970 by a number of specialists: Leo Oppenheim translated the texts, Robert Brill added chemical interpretations, and Dan Barag and Axel von Saldern put in additional sections on Mesopotamian glass vessels. The translations remain unchallenged to any significant extent, and the work remains monumental in its scale and accuracy. However, the interpretation of the texts, particularly by those interested in ancient glass technology, has been problematic. Moorey sums up the general mood when he says 'chemists interested in ancient glass technology have been heard to describe [the ancient texts] as "useless" for reconstructing ancient methods and recipes' (Moorey 1994). This chapter challenges this view, taking a fresh look at the translations in the light of the intervening 30 years of glass research.

The Tablets

In 1849, Sir Henry Layard and his team were excavating at the Palace of Sennacherib in the city of Nineveh, when they opened 'two large chambers of which the whole area was piled a foot or more deep in tablets' (Seton Lloyd 1955). Three years later, Layard's assistant Hormuz Rassam made a similar discovery on the same mound at the Palace of Sennacherib's grandson Assurbanipal. In total, some 25,000 tablets were recovered, many of which found their way eventually to the British Museum. Since most of these tablets actually date to the reign of Assurbanipal (668–627 B.C.E.), these finds have become known as the 'Library of Assurbanipal'.

The collection of all the tablets into one 'library' can be explained by reference to some of the texts themselves. One tablet contains the following instruction (Roux 1992):

> When you receive this letter [writes Assurbanipal to a certain Shaduna] take with you these three men [their names follow] and the learned men of the city of Barsippa, and seek out all the tablets [. . .] hunt for the valuable tablets which are in your archives and which do not exist in Assyria and send them to me [. . .] when you see any tablet or ritual about which I have not written to you, but which you perceive may be profitable for my palace, seek it out, pick it up, and send it to me. (Waterman 1930–36:vol. IV, 213, no. 6).

Several instructions very similar to the one above can be found in the texts, and Roux (1992:355) believes these represent a 'craving for culture' among the Assyrian kings, perhaps in an attempt to legitimise their political power and position.

The tablets fall into many different types of texts, the analysis of which is beyond the scope of this chapter. However, broadly speaking they can be divided into two main groups (Roux 1992): archive documents (royal inscriptions, letters, and administrative texts) and library documents (literary texts detailing myths and legends and scientific texts). Among the scientific texts, works on divination and exorcism predominate, but they also include what Oppenheim calls 'procedural instructions' (Oppenheim et al. 1970:5). These procedural instructions are written in the second-person singular and tend to be short, stereotyped, and technical in nature.

They detail methods for use in mathematics, rituals, and medical care, all of which would fit well within the role of the priestly/scribal authors of the texts (Robson 2001). However, in addition there are a few texts that appear to detail the chemical preparation of different materials, especially glass and perfume. Some authors (Moorey 1994; Oppenheim et al. 1970) have argued that these chemical texts do not fit well with the priestly/scribal role, in that they appear to give 'recipes' for the manual production of these materials, a role perhaps more fitted to an artisan than to a priest. However, Robson (2001) has pointed out that the cultural associations of glass were very different in the ancient world to those in the modern world. She has demonstrated that different colours of glass were viewed as having special magical properties that included beneficial and healing elements. These magical and medical fields very much lay within the priestly role, and therefore Robson finds it less surprising than some that 'recipes' for making such 'artificial stones' and perfumes are included within this corpus of literature.

The Glassmaking Texts

From the whole corpus of ancient cuneiform texts, about a dozen examples of procedural texts outlining the production of glass are known. Three of these date from the mid-2nd millennium B.C.E., one from Hattusha, one from Babylon, and one from an unknown location (Oppenheim et al. 1970:22–23). All the others were found in the Library of Assurbanipal, and Oppenheim judges these to be *n*th generation copies of Late Bronze Age originals (Oppenheim et al. 1970; Robson 2001), therefore, it is likely the original information for all the tablets dates from the same period, close to the beginning of glassmaking in the mid-2nd millennium.

Some of the texts are extremely simple and stylised, for example:

> For a mina of *zukû* glass: 10 shekels of lead, 15 shekels of copper, half (a shekel) of *anzahhu* glass, ½ (a shekel) of antimony: these are the ingredients of Assyrian red glass.
> For a mina of *zukû* glass: a sixth (of a mina) of lead, 14 (shekels) of copper, 10 shekels of antimony, a shekel of *anzahhu* glass: these are the ingredients of Akkadian red glass.
> (after Robson 2001:50 and Oppenheim et al. 1970:63–64)

Some are more descriptive:

> If you want to produce lapis lazuli-coloured glass, you grind separately 10 minas of *immanakku*-stone, 15 minas of *naga*-plant ashes, and 1½ minas of 'white plant'. You mix these together. You put them into a cold kiln which has four openings and arrange (the mixture) between the openings. You

burn a good and smokeless fire until the 'stone' becomes fritted. You take it out and cool it off. You grind it again. You collect (the powder) on a clean slab. You put it into a cold 'doorway' kiln. You burn a good and smokeless fire until it (the 'metal') becomes golden. You pour it onto a baked brick. Its name is *zukû*-glass – 'clear'.

You place 10 minas of [. . .] copper on a clean slab. You put it into a hot 'doorway' kiln and close the door of the kiln. You burn a strong and smokeless fire until the copper compound glows. You crush and grind 10 minas of *zukû*-glass. You open the door of the kiln and throw (the ground glass) upon the copper compound and close the door of the kiln again. Once the *zukû*-glass dissolves onto the copper compound you stir it once, twice, three times with a poker until you see a 'frost' (of molten glass) on the tip of the poker. If the 'stone' looks like ripe grapes you boil (it). You pour it onto a baked brick. Its name is *tersîtu* – 'preparation'.

You collect 10 minas of preparation, 10 minas of *bûsu*-glass, filtered glasswort ashes, not returned (?), ½ mina of mother of-pearl (?), 1⅔ minas of washed *anzahhu*-glass on a new slab. You put it into a kiln with 4 openings and place it on a support. The base of the slab must not touch the kiln. You burn a good and smokeless fire [. . .]. Once your mixture is melted you cool (it). You take (it) out and grind (it). You collect (it) on a clean slab. You put (it) into a cold 'doorway' kiln. You burn a good and smokeless fire. Not until the 'stone' glows do you close the kiln door. Once it is glowing you close the kiln door and stir it once in front of you until it becomes golden. After it has become golden you will see 'frost' (forming at the tip of the poker). If the 'stone' is solidifying you pour it onto a new slab and lapis lazuli-coloured glass will come out of the cool kiln. (after Robson 2001:51 and Oppenheim et al. 1970:32–34, Tablet A 13–42)

Yet others have strong ritual and magical associations:

> When you set up the foundation of a kiln for glass, you search in a favourable month for a propitious day so that you can set up the kiln's foundation. As soon as you have completed the kiln and built it, you install divine *kûbu*-images there. No outsider or stranger should enter; unclean persons should not pass in front of them. You regularly sprinkle strewn offerings before them. On the day when you place the 'metal' inside the kiln you make a sheep sacrifice before the divine *kûbu*-images. You place juniper incense on the censer; you make a fire in the bottom of the kiln and place the 'stone' in the kiln. The persons whom you allow to come near the kiln must be ritually clean so that you can let them come down to the kiln. The wood you burn in the bottom of the kiln should be thick, peeled poplar wood: logs without knots, bound together with leather straps and cut in the month of Abu (that is, July–August). This wood should go in the hearth of the kiln. (Oppenheim et al. 1970:32–34, Tablet A 13–42)

In summary, there are a number of different types of text, and many are duplicates. In all, there are about 40–60 different 'sections' or 'paragraphs', and the shorter formulae consist of just one of these sections, such as the formulae for Akkadian red glass and Assyrian red glass given above. The longer formulae run to several sections, such as the formula for lapis lazuli-coloured glass and the *kûbu*-images text above. All these texts present many challenges in translation and interpretation. They are often incomplete, areas of the tablet being damaged or missing, obscuring vital words and phrases. They are also full of rare or unique words, the meaning of which can only be guessed at from the context, making the danger of a circular argument (the translator looking to the glass technologist for the meaning of the word and the technologist looking to the translator for details of the raw materials for ancient glassmaking) ever present. In addition, the texts lack any pretence at a 'scientific' mode of presentation, and their original use or purpose is obscure, something that is discussed in more depth below. Finally, the recipes often describe the making of special types of glasses, the word for which appears to occur only in the glassmaking texts, for example, *zukû*-glass (see above). Since no finished object of these glasses is ever described, they have been interpreted (Oppenheim et al. 1970) as intermediate glasses – stages in the glassmaking process before the production of a final, coloured glass – which Oppenheim terms 'primary glasses'.

Replication: Brill's Experimental Work

Despite the problems described above, when the longer texts are read, they do appear to have a certain degree of technological logic to them. The production process follows a reasonably rational order. Where the words are known for the raw materials in use, they seem to indicate materials that one might expect could be included in the production of glass. Noticing these consistencies, Robert Brill attempted to replicate one of the glass texts in the laboratory (Oppenheim et al. 1970). The procedure followed was that detailed above for the production of *zukû*-glass and is laid out in flowchart form in Figure 3.1. Brill took 10 parts by weight *immanakku*-stone, translated by Oppenheim as quartz pebbles, and added 15 parts of plant ash from a desert plant. The third component, one and a half parts of 'white plant', could not easily be interpreted but may have been a lime rich plant or other material. Either way, since it was a relatively minor component of the system, Brill left it out of his experimental replication. The mixture was fired and produced a pale amber glass 'of unexpectedly high quality' (Oppenheim et al. 1970). Brill therefore demonstrated that the tablet apparently describing the production of *zukû*-glass did describe a feasible method for the production of a pale-coloured glass. Unfortunately, owing to the fragmentary nature and questionable interpretations of the other tablets, Brill was unable to follow any other potential 'recipe', so the production of *zukû*-glass remains the solitary example.

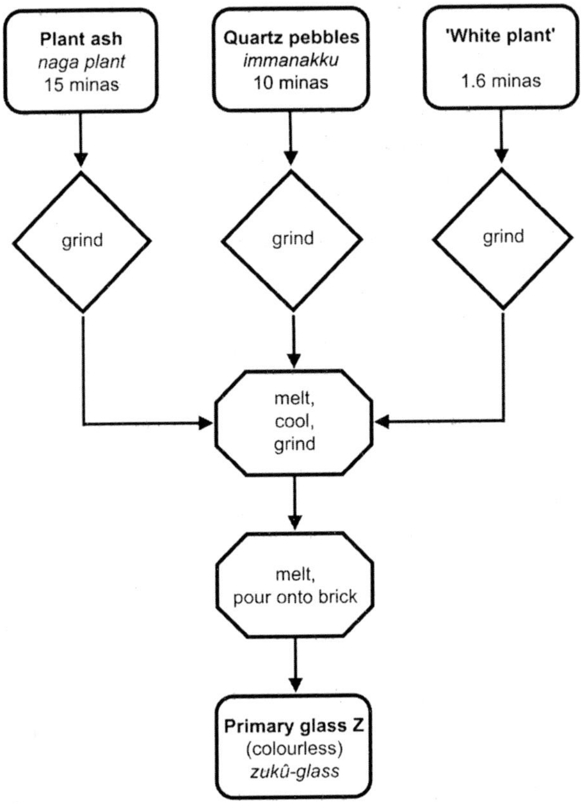

Figure 3.1 • Flowchart of Tablets A and B, Section 1: *zukû*-glass (Oppenheim et al. 1970:32–34).

A Question of Use

The most fundamental problem for the interpretation of these texts is the question as to what the original writers intended them to be used for. Their intended use dictates the degree of access to technological information that the original writers would have desired or required and the accuracy with which this technological information would need to be recorded and hence transmitted. Moorey (1994) has stated his belief that the texts 'served literary or administrative purposes, they did not represent a body of technical or scientific literature'. He goes on to say that 'these texts are part of a traditional Mesopotamian literary enterprise: the compilation of lexical lists [. . .] [and] concern themselves with little more than the appropriate technical vocabulary [. . .]' Therefore, in Moorey's view these texts are lexical lists, essentially dictionaries, strung together with joining words. They are therefore of little use in the construction of ancient recipes for glassmaking, since that was not their purpose; therefore the details they record do not lend themselves to that role.

However, Robson (2001) has highlighted ritual and magical aspects of the texts, something that is especially evident in the ritual text concerning the *kûbu*-images, given above. This work clearly shows that some of the texts contain a ritual element, which appears to be recording actual rituals that were to take place around the kiln site. This would be very much the realm of the priest/scribe, and therefore one can reasonably assume that these represent at least some of the actual ritual activities that may have gone on around the production of glass.

Similarly, as detailed above, Brill has demonstrated in his replication work that there was some practical value to some of the texts. Now it could be entirely by chance that the weights for the production of *zukû*-glass do actually make a satisfactory glass, but perhaps it is more likely that the texts do record something of the practical information of glassmaking.

The question, therefore, is how can this be resolved? Are these texts lexical, ritual, practical, or some combination of these types? Robson has simplified matters a little. She says: 'The boundaries between science and religion, medicine and magic were always blurred in the ancient Near East: the spiritual was inseparable from the rational' (Robson 2001:54). In other words, in the ancient mind, the ritual and the practical were as one and were inseparable. Both are vitally important to the task in hand (in this case the manufacture of glass), and breaking a ritual taboo was considered as serious to the finished result as adding too much of one raw material or another. The 'ritualising' of glassmaking has the effect of stifling technological choice and prohibiting experimentation and innovation. The laying down of mechanistic rules that should be rigidly followed is a classic symptom of such ritualisation (see Shortland 2007). The use of ritual in a manufacturing process should not be regarded as a useless, counterproductive interloper. It can have many beneficial effects, but the one area that it will prevent is *technological change*, and it could be argued that the great conservatism seen in the production technology of these early glasses is some of the best evidence for the ritualisation of the methods used.

However, it is also perhaps true that the use of the text has changed through its history. The information recorded in the mid-2nd millennium B.C.E. may well have had little or no practical or even ritual use by the mid-1st millennium. By this time, its use could well have been almost entirely lexical, part of the traditional Mesopotamian literary enterprise that Moorey writes about. Hence, the use of the text seems to have evolved, taking in elements of ritual/practical and lexical at different times in its history. However, the question remains, can these mid-1st millennium B.C.E. lexical texts say anything about the technology and technological processes of the mid-2nd millennium B.C.E. glass industry?

Texts and Technology

To investigate whether any technological information remains in the texts, it is necessary to question them in some detail. Moreover, the actual questions used must reflect the fragmentary nature of both the tablets and the translations. Hence, since whole 'recipes' cannot be verified because of our modern incomplete understanding of the text, it is not sensible to condemn them as 'useless' for failing to give all the details for the production of each of the different glasses. It is far more instructive to break the 'recipes' down into more manageable and intelligible units and investigate their use of individual raw materials and production methods.

Looking through all the tablets, where translations are possible, one sees that all the raw materials mentioned are the sort of materials that one might expect would be used in the production of glass. There are no inconsistencies or strange materials that might lead one to expect that there are fanciful additions being made. Indeed, some raw materials are quite surprisingly accurate in their citing. An example of this can be found in Tablets A and B, Section 8 (Oppenheim et al. 1970), describing the production of lapis lazuli-coloured glass. One of the raw materials mentioned in the production of this glass is four[1] shekels of 'red alum' (*ga-bi-e samu*). When Oppenheim was translating this passage in the late 1960s, this may well have appeared to be an odd material to be putting into the glass, and it is interesting to note that he does not make any mention of its place here in his commentary on the text. However, the work of Alex Kaczmarczyk in the mid 1980s (Kaczmarczyk 1986) has shown that pink alums from the Western Oases of Egypt were the source of cobalt colourant for much if not most of the cobalt-coloured, lapis lazuli-imitating glass of the mid-2nd millennium B.C.E., certainly in Egypt and perhaps further afield as well. The texts were therefore hinting at the presence of alum in the production of some lapis lazuli-coloured glasses long before analysis showed that this was the probable source.

Furthermore, the texts contain excellent descriptions of processing methodologies. Perhaps the best example of this is in Tablet A, Section 4 (Oppenheim et al. 1970):

> If you want to produce *zagindurû*-coloured glass, you grind finely, separately, 10 minas of *immanakku*-stone, 12 minas of plant (ashes). You mix (these) together, you put (the mixture) into a cold kiln which has four fire

> openings and arrange (the mixture) in between the four openings. You keep a good and smokeless fire burning. As soon as your mixture glows red, you take it out into the open air and allow it to cool off. You grind it finely again [. . .]

Here we have a description of a process, with much incidental detail. The plant ash and quartz pebbles are ground and well mixed. They are put in a cold kiln, spread out well and heated in a smokeless atmosphere until red hot. They are then rapidly cooled in air and ground again. This is an exact description of a fritting process, one of the intermediate stages in the production of glass (Turner 1956). It is very unlikely that such an accurate description could have been derived by chance, it must reflect some technological knowledge remaining in the text, despite the millennium of copying that it has undergone. The texts should therefore be taken seriously as a potential resource for technological information, so long as their peculiarities and complex history are taken into account.

Glassmaking Methodology: A 'Rational' Process?

Given that the tablets do seem to have at least some accurate technological information, there is a further point that must be raised from the texts – the presence, indeed, preponderance of texts that appear to deal with the manufacture of intermediate or 'primary' glasses. This raises the very important issue of how many intermediate stages were involved in the glassmaking process.

Modern glassmaking has strongly influenced the way ancient glassmaking has usually been interpreted. Ancient glassmaking of this period is usually thought (Henderson 2000; Nicholson and Henderson 2000; Turner 1956) to use a very rational, linear process such as that shown in the flowchart in Figure 3.2a. The raw materials are heated together, perhaps via a fritting stage, and the glass is formed. The process is efficient and simple. However, these texts suggest a more complex approach. The text concerning the manufacture of *zukû*-glass given above is a good example. Although the first half of this text is clearly drawn out in the flowchart (Figure 3.1) and replicated in Brill's work, this is only the first part in the process to produce the lapis lazuli-coloured *zagindurû*-glass of the first line. The full flowchart for the production of this glass is shown in Figure 3.3. This shows that the primary *zukû*-glass is mixed with 'slow copper' (a colourant) and then at a later stage more *zukû*-glass and two other primary glasses are added, in addition to as much plant ash 'as is needed'. In simplified terms, Figure 3.2b shows the flowchart for such a production methodology. Note that, from modern analysis of only the raw materials and the final glass, the intermediate stages would probably be invisible to the student of ancient glass technology. Our being able to analyse only the beginning and the end of the process would automatically lead to the process being reconstructed not in terms of the multiple transitional stages of Figure 3.2b but in terms of the simple linear approach of Figure 3.2a, since the intermediate stages are analytically invisible in the end product. There are only two ways that the presence of these intermediate stages could be hypothesised: a literary or iconographic source, such as the glassmaking texts discussed here, or the presence of one or more of these intermediate stages at some ancient glassmaking site. It is just possible that the latter may have been discovered at Amarna and may confirm the interpretations of the former.

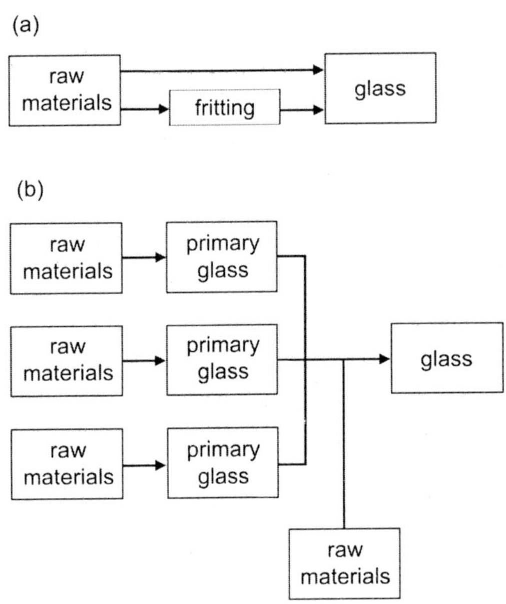

Figure 3.2 • Methodologies for glassmaking: (a) the rational, modern approach; (b) the approach implied by the glassmaking texts.

The Amarna Frits

The glassmaking sites at Amarna in Middle Egypt date from the mid-14th century B.C.E. and were first excavated by W. M. F. Petrie in the late 19th century (Petrie 1894). Further work on the site has been carried out recently by Paul Nicholson (Nicholson 1995), and the site remains one of the best studied of the ancient glassmaking sites. One of the most unusual materials found at the site by both Petrie and Nicholson were masses

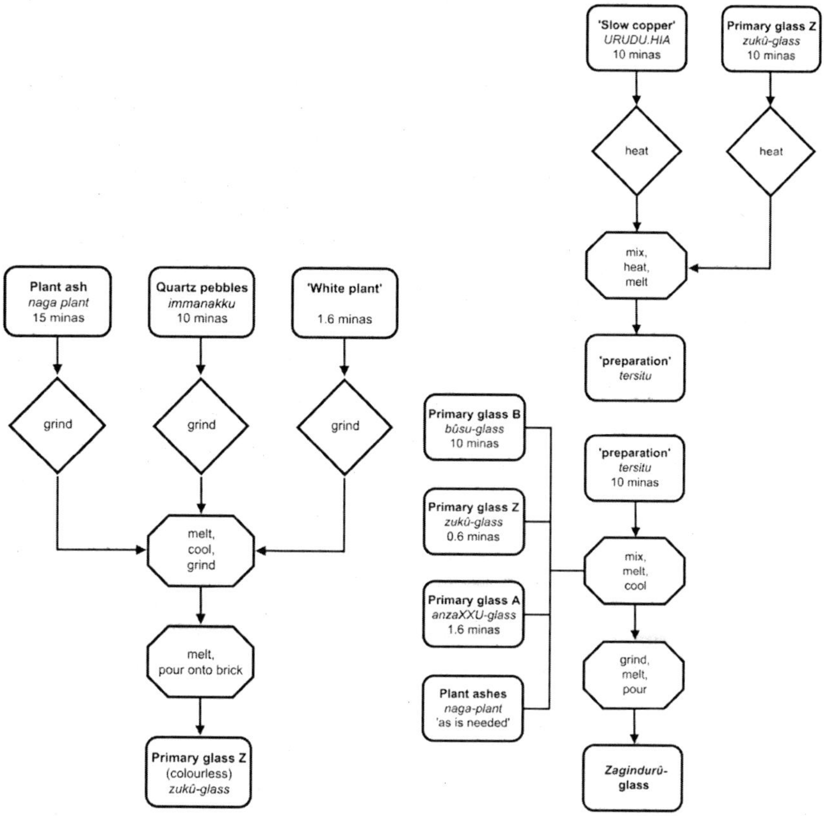

Figure 3.3 • Flowchart of Tablets A and B, Sections 1–3: lapis lazuli-coloured, *zagindurû*-glass (Oppenheim et al. 1970:32–34 and Robson 2001:51).

of vitreous, blue material described by them both as 'frit' (Figure 3.4). This frit consists of remnant quartz grains in a glassy matrix coloured by cobalt or occasionally copper. This frit was first considered to be a simple glassmaking frit, the remnant of the fritting stage described in the fritting text above and the simple flowchart of Figure 3.2a (Petrie 1909; Turner 1956). However, the situation is more complex, since it was found that the composition of the frit made it impossible simply to melt it and create a glass of the correct ancient composition (Shortland 2000). Other interpretations of the frit have been attempted, including the idea that it was part of the faience-making process rather than glass (Shortland 2000), but none of them has been entirely satisfactory. Detailed discussion of the place of the frits is beyond the scope of this chapter (but see Tite and Shortland 2003), but there is a further possibility. It could be that this

Figure 3.4 • Amarna 'frit' sitting in a 'fritting pan': (a) cross-section; (b) from above.

'frit' is actually one of the primary glasses talked about in the texts. This would certainly explain why it does not have a composition intermediate between the raw materials and glass. If it were a primary glass, other materials would have been added to it to create the final glass, perhaps including other primary glasses and raw materials. It would therefore not need to have the intermediate composition. It may be, therefore, that the Amarna frits are primary glasses and support the interpretation of the texts that suggests a multiple-stage production technique.

The question then remains as to why use multiple stages when one stage is probably possible. To suggest an answer to this one must go back to the statement that the ritual and practical were inseparable. Even in modern laboratories, ritual is involved in the processes carried out. Often in a routine process, the preparation of samples for archaeometric dating, for example, routines become incorporated because they seem to aid the process, although exactly why is sometimes not at all clear. Other routines are handed down, technician to technician, sometimes without including the real reasoning behind why they are actually done – they are done because they have always been done, and they work. In consequence, even in a modern laboratory, there is often a great reluctance to remove stages from a process that to all intents and purposes is working. Therefore routines tend to agglomerate and remain. This is in a modern laboratory, where spiritual and theological concerns are minimal. In the ancient situation in which the ritual and practical are essentially the same, the tendency to incorporate stages and reluctance to remove them will have been far greater. Random failures in the kiln will lead to new ritual/practical stages being added and then retained to ward off further failures. The multistage flowchart in Figure 3.2b therefore looks much more reasonable from an ancient point of view than does the modern linear progression of Figure 3.2a.

It has also been speculated (P. Vandiver, pers. comm.) that there may be practical values in having intermediate stages. In the ancient world, where raw materials are variable and of questionable quality, there may be virtue in adding intermediate stages in the process. This is because fluctuations in raw material quality are likely to lead to failures in the melt. If these failures are identified early (at an intermediate stage) they can be discarded without affecting the main mass of the melt, hence sparing a lot of work, fuel, and potentially valuable raw materials. Raw material variation in the linear model would result in the discard of the batch and the requirement to start all over again. In the multistage model, only the failed fraction of the material has to be discarded. There are, therefore, perhaps positive practical outcomes to using intermediate primary glasses, but it should be stressed that the *reasons* why they were used are almost certainly dominated by *ritual* considerations. The practical outcomes are secondary, emergent phenomena and may not have been apparent to the workers involved.

Conclusion

This chapter has attempted to show that despite their complex lexical history, the cuneiform glassmaking texts do seem to have interesting and useful ritual and practical information contained within them. It is true that owing to damage to the tablets and problematic words the process of interpretation is difficult, and it is perhaps not surprising that it is not possible to follow the full 'recipe' for the production of glass in any single case. Yet, if one examines not the full texts but individual parts of the text, one clearly sees that important information concerning methods and raw materials is there; hence practical information is preserved. This is especially true concerning the overall method of glass production, the texts suggesting a multistage process with intermediate primary glasses being used as opposed to the simpler linear model. It is possible that the frits found on the glassmaking site of Amarna in Egypt may be an example of such an intermediate primary glass. In summary, it certainly seems clear that the texts do have their place in the interpretation of ancient glassmaking technology and do deserve further investigation in light of the 30 years of progress in the subject.

Note

1. Perhaps an error in Oppenheim's translation; there may in fact be five.

References

Henderson, J. 2000. *The Science and Archaeology of Materials.* London: Routledge.

Kaczmarczyk, A. 1986. 'The source of cobalt in ancient Egyptian pigments', in J. S. Olin and M. J. Blackman (eds.), *Proceedings of the 24th International Archaeometry Symposium.* Washington, DC: Smithsonian Publications, 369–76.

Moorey, P. R. S. 1994. *Ancient Mesopotamian Materials and Industries.* Oxford: Oxford University Press.

Nicholson, P. T. 1995. 'Glass-making and glass-working at Amarna: Some new work', *Glass Technology* 37(1):11–19.

Nicholson, P. T., and J. Henderson. 2000. 'Glass', in P. T. Nicholson and I. Shaw (eds.), *Ancient Egyptian Materials and Technology.* Cambridge: Cambridge University Press, 195–226.

Oppenheim, A. L., R. H. Brill, A. von Saldern, and D. Barag. 1970. *Glass and Glass-Making in Ancient Mesopotamia.* New York: Corning Museum of Glass.

Petrie, W. M. F. 1894. *Tell el-Amarna.* London: Methuen.

———. 1909. *The Arts and Crafts of Ancient Egypt.* London: T. N. Foullis.

Robson, E. 2001. 'Technology in society: Three textual case studies from Late Bronze age Mesopotamia', in A. J. Shortland (ed.), *The social Context of Technological Change: Egypt and the Near East 1650–1550 BC.* Oxford: Oxbow Books, 39–57.

Roux, G. 1992. *Ancient Iraq.* London: Penguin.

Seton Lloyd, I. 1955. *Foundations in the Dust.* London: Penguin.

Shortland, A. J. 2000. *Vitreous Materials at Amarna: The Production of Glass and Faience in 18th Dynasty Egypt*. BAR International Series, S827. Oxford: Archaeopress.

———. 2007. 'Who were the glassmakers? Status, theory and method in mid-second millennium glass production'. *Oxford Journal of Archaeology* 26(3), 261–74.

Tite, M. S., and A. J. Shortland. 2003. 'Production technology for copper and cobalt-blue vitreous materials from the New Kingdom site of Amarna: A reappraisal', *Archaeometry* 45(2):285–312.

Turner, W. E. S. 1956. 'Studies in ancient glasses and glass making processes. Part V. Raw materials and melting processes', *Journal of the Society of Glass Technology* 40:277–300.

Waterman, L. 1930–1936. *Royal Correspondence of the Assyrian Empire*. (4 vols.). Ann Arbor: University of Michigan Press.

CHAPTER 4

Pliny on Roman Glassmaking

IAN C. FREESTONE

ABSTRACT *Pliny's* Natural History, *dedicated in 77* C.E., *some two years before his death, is the major written source on glassmaking from Antiquity. It underpins much of our current understanding of Roman glassmaking materials, in particular the use of Egyptian soda as a flux and of the sands of the Levantine coast as a source of silica. However, from the perspective of modern glass technology, Pliny's account has been found confusing and ambiguous, especially in the descriptions of technological processes. Recent archaeological and archaeometric work allows clarification of a number of the major ambiguities. The stages of melting indicated by Pliny may be equated with the subdivision of production between primary and secondary workshops, recognised from excavation and the analysis of glass. Constraints may be placed on the identification of additives such as magnes lapis and lustrous stones. An intriguing issue remains Pliny's reference to the use of glassmaking sand from the coast near Naples, for, although the chemical evidence is largely negative, certain compositional trends shown by the sands are repeated in some Roman glass. It is doubtful if either scientific archaeology or a reading of Pliny alone would have yielded a fully adequate explanation of Roman glass technology.*

Introduction

Most students of ancient glass technology turn at some point to the pages of Pliny's *Naturalis Historiae*, in the hope of some insight into ancient practices. One suspects that many are in turn enlightened, intrigued, and eventually frustrated, sharing the view of W. E. S. Turner, in his pioneering study of early glass technology, that some of the entries that are potentially most significant are 'obscure' (Turner 1956b:293). Even so, Pliny is undoubtedly an important source of reference for students of early glass and is routinely referred to in the literature. Indeed, a German translation has been produced specifically for glass technologists (Knoll et al. 1979). However, there remains a good deal of uncertainty about Pliny's account.

The *Natural History* provides the major written source on glass production in Antiquity. Dedicated to Titus in 77 C.E., about two years before Pliny's death in the eruption of Vesuvius, it comprises 37 books and incorporates, according to Pliny himself, some 20,000 facts (*Nat. Hist.* Preface:17). It combines the knowledge of earlier Greek and Roman science with contemporary accounts, some no doubt second hand and others clearly based on Pliny's own observations. A small part of only one of these, Book 36, is devoted to glass. Pertinent information is to be found elsewhere, however. In particular, part of Book 31 describes soda, its occurrence and production. The present account depends primarily on the translation of Eicholz (1962), and quotes are from this source unless stated otherwise. In addition, the interpretation draws on the accounts of Healy (1981, 1991, 1999) and Knoll and colleagues (1979) at various points.

An account of Pliny the man, his background and outlook, is provided by Healy (1999), based in part on the letters of Pliny's nephew, Pliny the Younger. This ground will not be revisited beyond summarising a few points concerning the sources of information in the *Natural History.* According to Pliny (Healy 1991, *Nat. Hist.* Preface:17), he compiled his information from 2,000 volumes by 100 authors; Healy (1981) records 146 Roman and 327 foreign authors. In addition, Pliny derived his information from local people, Roman commanders, provincial governors, and other people with whom he was able to talk and correspond. Pliny himself travelled outside Italy, and some sections are therefore likely to be based on personal observation. He held a military command in Germany, and the evaluation by Healy (1999) supports the view that he held a procuratorship in Hispania Tarraconensis with a possible further one in Africa. In Italy itself, Pliny probably travelled quite widely. Born in northern Italy in 23 or 24 C.E., he moved to Rome in the 30s, and later in life became admiral of the fleet at Misenum, Bay of Naples.

Previous commentators have adopted a more or less similar approach to Pliny's chapters on glass, underpinning their interpretations by an understanding of postmedieval and modern glassmaking practices. The time is now right for a reexamination of his account, in the context of the extensive archaeological information that has recently become available on the composition and technology of Roman glass and glass made in the Roman tradition.

This chapter focuses on a number of key issues in Pliny's account that relate to the raw materials and organisation of the production of the glass material itself, as opposed to the fabrication of artefacts of glass. The much-quoted but apocryphal tale concerning the discovery of glass will not be revisited, and anecdotes concerning the role or status of glass in society are best considered elsewhere. The fabrication of vessels falls outside the scope of this chapter, but the colouration of glass is included. Before turning to Pliny's account of glass technology, I provide an outline of the current understanding of Roman glass production and its organisation, which derives largely from recent archaeological, archaeometric, and ethnoarchaeological studies. This provides a context that clarifies the meaning of Pliny's text in several key respects.

The Organisation of Glass Production

The model of ancient glassmaking that held sway in the second half of the 20th century was summarised by Turner:

> The melting process was carried out in two stages, first a comparatively low temperature operation where the sand and the ash (or soda) were converted to a frit, the second in which the frit was converted to glass in crucibles in a melting furnace. Flinders Petrie considered that this two-stage process was carried out in 18th-Dynasty Egypt. From Pliny's obscure account the process in his time would seem to have involved three stages [. . .] (Turner 1956b:292)

This interpretation, involving the production at relatively low temperatures of a partially reacted but incompletely melted alkaline silicate frit and then fully melting it in crucibles, appears to have been based largely on postmedieval European glassmaking practices and the 17th-century account of Antonio Neri, as translated and amended by Christopher Merret (republished by Cable 2002). Support for it was found by Turner (1956b) in the early 12th-century writings of Theophilus. It is now clear that it is misleading in a number of key respects, at least for the Greco-Roman and early Islamic worlds, from the late 1st millennium B.C.E. through to the 9th century C.E.

Views of ancient glassmaking have modified dramatically following the discoveries of large workshops in Egypt (Nenna, Vichy, and Picon 1997; Nenna, Picon, and Vichy 2000) and the Levant (Aldsworth et al. 2002; Gorin-Rosen 1995, 2000; Tal et al. 2004) that had the prime function of making glass from its raw materials, without necessarily making vessels. These workshops specialised in melting large batches of material in a single firing, as exemplified by the 8-tonne slab of failed glass at Bet

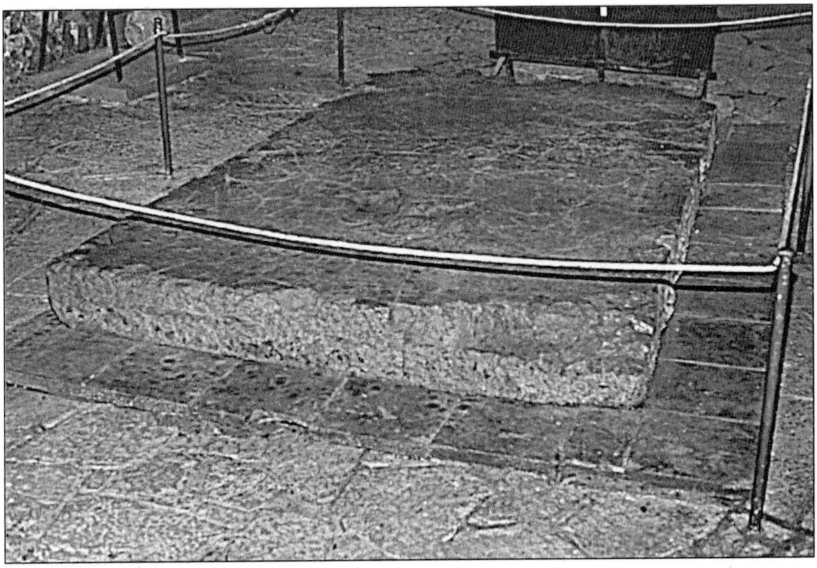

Figure 4.1 • The slab of failed glass at Bet She'arim, Israel; dimensions of slab are 3.40 × 1.95 × 0.45 m (photo I. C. Freestone, digitally enhanced by A. Milton).

She'arim, Israel (Figure 4.1; Brill and Wosinski 1965). The slab appears to have been the result of a well-established and widespread procedure that had failed owing to the use of incompatible raw materials (Freestone and Gorin-Rosen 1999). This particular slab is thought to date to the early Islamic period (8th–9th centuries C.E.), but there is also evidence of large-scale melting in earlier periods (for example, Nenna et al. 2000). After melting, such large slabs of glass were broken up into angular lumps (chunks) and distributed to secondary workshops, where they were remelted and fabricated into vessels. Analysis of chunk glass from wrecks of the Roman period has demonstrated the presence of a limited number of glass groups that are likely to correspond to an equivalent number of primary glass production centres that dominated glass production at that time (Foy, Vichy, and Picon 2000; Foy, Picon, and Vichy 2000; Foy et al. 2003). In addition, analyses of late Roman (5th–7th century) vessels from the eastern Mediterranean demonstrate that the glass itself is likely to have been made in a limited number of primary workshops or factories (Foy et al. 2003; Freestone, Greenwood, and Gorin-Rosen 2002; Freestone, Ponting, and Hughes 2002).

Ethnographic corroboration for the production of large slabs of glass using traditional furnace technologies comes from northern India, where Kock and Sode (1994; Sode and Kock 2001) have investigated an apparent continuation of this approach. Here, sand and alkali are fused into slabs of glass weighing some 3 tonnes, in a process that attains relatively low temperatures (estimated at around 900°C) but that continues for 30 days. A feature of this process is that an initial fritting, in the sense of a low-temperature sintering of sand and alkali into a semivitrified mass, is not carried out. The sand is initially roasted in the absence of added alkali, the purpose apparently being to remove organic matter to allow the production of an oxidised glass (Sode and Kock 2001).

Pliny indicates that, for many years, glassmaking had depended on a mixture of soda with the sand from a beach near the mouth of the River Belus, on the coast of Syria:

> This is supposed to be the source of the River Belus, which after traversing a distance of 5 miles flows into the sea near the colony of Ptolemais [. . .] The beach stretches for not more than half a mile, and yet for many centuries the production of glass depended on this area alone [. . .] (*Nat. Hist.* XXXVI:190)

He then goes on to describe how glass was made:

> Glass, like copper, is smelted in a series of furnaces, and dull black lumps are formed [. . .] After being reduced to lumps, the glass is again fused in the workshop and is tinted. Some of it is shaped by blowing, some machined on a lathe and some chased like silver. (*Nat. Hist.* XXXVI:193)

Pliny here indicates two stages of melting. Healy (1999:354) has argued that Pliny used 'series' here to conflate several processes, as he appears to have done elsewhere when describing metallurgical activities. It is also possible, in the light of the archaeological evidence, that the 'series' of furnaces is perhaps analogous to the array of 17 tank furnaces for melting raw glass found at Bet Eli'ezer, near Hadera in Israel (Gorin-Rosen 1995, 2000). The glass slab formed in such a furnace might well appear dull and opaque, particularly if it was reduced to the point where the ferri-sulphide complex formed and produced a deep olive glass (Healy 1999; cf Schreurs and Brill 1984). It was broken up into lumps and taken to a separate workshop where it was remelted, coloured if desired, and worked into vessels. The paragraph may be interpreted to indicate that the glassmaking furnaces were at a separate location from the workshop, where vessels were fabricated; this would be consistent with current understanding. Pliny gives a strong indication that this was his meaning by the analogy with the production of copper, which was traditionally smelted from raw materials in furnaces near the mines and transported over long distances as ingots to workshops, where it was alloyed and fabricated into artefacts.

There is no indication in Pliny's account of the production of a crystalline or semivitreous low-temperature frit as a first stage in glass production. The primary smelting stage (that is, the melting of the raw materials) produced a glass, and lumps of raw glass were taken to the workshops to be fabricated into vessels, as is fully consistent with our current understanding of Roman glassmaking.

Pliny states that the method of glassmaking described above was

> the old method of making glass. Now, however, in Italy, too, a white sand which forms in the River Volturno is found along six miles of the seashore between Cuma and Literno. Wherever it is softest, it is taken to be ground in a mortar or mill. Then it is mixed with three parts soda, either by weight or measure, and after being fused is taken in its molten state to other furnaces. There it forms a lump known in Greek as 'sand-soda' [*hammonitrum*]. This is again melted and forms a pure glass, and is indeed a lump of colourless glass. (*Nat. Hist.* XXXVI:194)

Just what it is about the first method that Pliny considers 'old' is somewhat ambiguous. The sense of the text appears to be that the new method uses different sand, from Italy, rather than the old sand from Syria-Palestine. Again it is fused with soda, directly to a glass. It is then taken to other furnaces (probably in other glass workshops) to be remelted (it must be assumed that 'in its molten state' refers to cooled glass, rather than a melt at 1000°C). The meaning of *hammonitrum*, a Greek word that translates as 'sand-soda', is not entirely clear. From its position in the text, it might indicate that when the material was remelted it formed the 'sand-soda'. However, it is unlikely that a previously molten material could be remelted to produce something more akin to the original sand and soda. More likely, 'sand-soda' was the term used by the craftsmen who carried out the second melting to describe the product of the first. If so, this passage may be interpreted as describing two stages of fusion, in primary and secondary workshops, as has been inferred for the earlier process, when using the Belus sand. Only two stages of melting appear to be unambiguously mentioned in this section.

Might the term *hammonitrum* refer to a partially crystalline frit, in the sense of 'a calcined mixture of sand and fluxes as material for glassmaking' (Thompson 1995:542)? Although the reading of the term is suggestive, at the present time the archaeology does not appear to support such an interpretation. The excavation of glassmaking installations that made glass in the Roman tradition has yielded fragments of raw glass, which have been completely molten, but a dearth of partly crystalline material. There is evidence for such an intermediate material from New Kingdom Egypt (Rehren and Pusch 2005; Shortland 2000 and this volume), but the raw materials, the scale of production, and the overall technological context were very different. In addition, Sode and Kock (2001) did not observe production of a frit in their study of similar glassmaking methods in modern India.

The glass production process of Pliny's time may therefore be read as being similar in both Syria and Italy in that it involved the blending of soda and sand and their melting in a furnace to form a primary glass, or *hammonitrum*. It was then taken to 'the workshop' or 'other furnaces', remelted, and used to fabricate vessels.

It is significant that in the 'old' method, the glass was 'tinted' in the workshop, whereas in the 'new' method it is remelted to form a 'colourless glass'. The emphasis on colourless glass in the 'new' method reflects the increasing fashion for colourless glass in the third quarter of the 1st century C.E., when Pliny was writing (see, for example, Price and Cottam 1998:15).

Raw Material Sources

Glass Chemistry

Virtually all Roman glass is soda-lime-silica glass, comprising three major components. Like other ancient glass and most modern glass, it is based on the chemical compound silica, silicon dioxide, SiO_2. Silica occurs widely in nature as the mineral quartz, which can occur as pebbles and is a major component of many sands. Very high temperatures, of around 1700°C, are required to melt pure quartz, and these were well beyond the reach of preindustrial technology. To make glass, it was necessary to add a flux that lowered the melting temperature of the silica to a more readily attainable range. The flux used by most glassmakers was (and still is) soda, sodium oxide, Na_2O. Soda does not occur naturally in its pure form but combined in salts such as sodium carbonate. When heated with silica, the carbonate is lost to the atmosphere as carbon dioxide gas.

Unfortunately, simple soda-silica glasses are readily corroded by water and may be completely soluble. To prevent this, a third component, known as a stabiliser, is required. The main stabiliser in most ancient glasses was lime, calcium oxide, CaO. Lime is found naturally in the form of limestone and marine shell, which are forms of calcium carbonate.

The secret of making a good soda-lime-silica glass lay partly in obtaining raw materials that provided the three components, soda, lime, and silica, in about the right proportions. Relatively minor amounts of impurities, such as aluminium and iron oxides, could be incidentally included in the batch as impurities, especially in the glassmaking sand. In its reduced state, a few tenths of 1% of iron impart a distinctive blue colour to glass, so careful selection of the sand used was essential.

In addition to the minerals mentioned, the ashes produced by burning some plants from the Mediterranean and Near East are rich in soda and lime and were widely used as glass fluxes in some periods – for example, in the 2nd millennium B.C.E. In general, the chemical compositions of glasses made with plant ash are readily distinguished from those of glasses made using mineral soda. Plant ash is a complex material, and, in addition to soda and lime, it contains significant quantities of potassium and magnesium oxides. These are carried through the melting process into the glass and provide a distinctive signature for plant ash glass that contrasts with that for glasses made using relatively pure sodium carbonate minerals where potassium and magnesium are present in much lower concentrations (Sayre and Smith 1967).

Soda

Pliny is quite specific about the source of soda used in glassmaking, *preferably soda from Egypt* (*Nat. Hist.* XXXVI:193). Just how this soda was gathered and prepared is less clear, however. Natural deposits of soda are well known from the Wadi el Natrun, some 50 km northwest of modern Cairo. Here the sodium bicarbonate mineral *trona* is among the assemblage that crystallises owing to the evaporation of a series of lakes that fill in the winter months (Shortland 2004). The crystallising trona crystals may be gathered from the edge of the lakes in the Spring, or trona may possibly be broken out as lumps from the massive material formed as the lakes dry. Pliny comments that soda sources in Egypt include those in the vicinities of Naucratis and Memphis. These locations have been discussed in some detail by Lucas and Harris (1962:265–66), who conclude that the Naucratis occurrence probably corresponds to the smaller natron lakes at Barnugi (modern al-Barnuj; Figure 4.2), whereas Memphis is most likely to correspond to the deposits of the Wadi Natrun.

Pliny also comments that soda 'of an inferior quality' may be extracted in 'the same manner as is salt, except that they pour sea-water into the salt-beds but Nile water into the soda-beds' (*Nat. Hist.* XXXI:109; Jones 1963).

Turner discussed in a literal sense the parallel that Pliny draws with the evaporation of seawater to yield salt, but noted that the yield of salts from the Nile would have been very low (Turner 1956b). Sayre and Smith (1967) observed that the relative alkali and alkaline earth contents of Nile water are more or less appropriate for glassmaking, although it is noted here that the high magnesium, potash, and iron contents would be inappropriate for Roman glass (see below). It is also possible that Pliny's description refers to the letting of Nile water into pits and its controlled evaporation to deliberately induce the efflorescence of salts in the soil, which could then be collected.

Figure 4.2 • Trona precipitating on the edge of a soda lake at al-Barnuj, Nile Delta, Spring 2004 (photo I. C. Freestone).

Pliny also indicates other sources of soda, including Thrace, and a Macedonian source, recently investigated by Ignatiadou and colleagues (2005). However, he does not mention these in the context of making glass.

Compositional analysis of ancient glass confirms that the source of alkali in virtually all Roman glass contained low potash and magnesia and was therefore naturally occurring soda, rather than the plant ash utilised by Bronze Age glassmakers and later by Islamic and Venetian craftsmen (for example, Henderson 1985; Sayre and Smith 1967; see above). Although analytical approaches to the determination of the geographical location of soda sources have not yet been devised, they may be forthcoming. For example, elements such as lithium and boron and their isotopes, which are likely to have derived mainly from the alkali used, might vary systematically between sources. Unfortunately, precise measurements of these components in ancient glass have not been made on any scale.

The Belus Sand

Pliny is one of a number of classical authors who cite the beach near the mouth of the River Belus, which meets the sea near Ptolemais (the modern River Naaman, running in to the Bay of Haifa, and the city of Akko, respectively), as a source of good glassmaking sand (*Nat. Hist.* XXXVI:190; see also Strabo *Geography* 16.2.25). Excavations in Israel have yielded abundant evidence of glassmaking and glassworking from the Roman through to the early Islamic period (Gorin-Rosen 2000), presumably exploiting the renowned glassmaking sands of the region.

A major step forward in our understanding of the use of the Belus sand was made by Brill (1988), who showed that the elemental and oxygen isotope composition of 4th-century Roman glass from the workshop at Jalame, Israel, could be closely replicated by mixing Egyptian natron with sand collected from the beach, near the mouth of the river. It appears that the calcium carbonate content of the Belus sand, present mainly as fragments of beach shell with subordinate limestone, is such as to provide all of the lime (calcium oxide, CaO) present in the glasses.

Extensive sedimentological investigations of Israel beach sands were carried out some years ago by Emery and Neev (1960). From these, some insight into the Belus sand and its suitability for glassmaking can be gained. The sands of the Israeli coast are derived ultimately from the Nile and are moved up the coast by the prevailing Mediterranean current and longshore drift. (For a general sedimentary model, see Goldsmith and Golik [1980] and Stanley and associates [1997].) Emery and Neev (1960) determined the calcium carbonate contents of beach sands and their colours for a series of beach stations spanning the coast of Israel. Their data are plotted in Figure 4.3 as two curves, showing weight

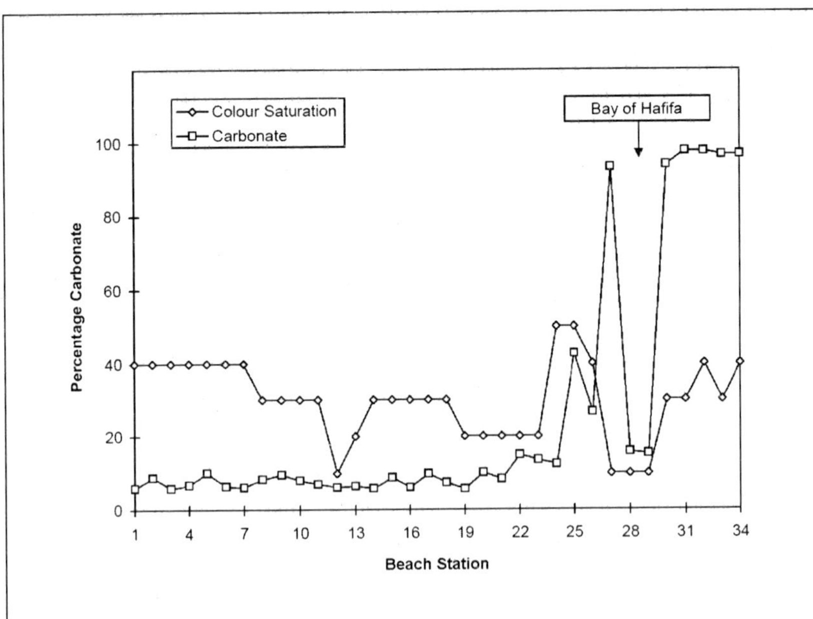

Figure 4.3 • Graph showing the variation in composition of beach sands on the Mediterranean coast of present-day Israel; carbonate in weight percentage and colour density in arbitrary units (data from Emery and Neev 1960).

percentage $CaCO_3$ and Munsell colour density versus location (beach station). The lower the colour density, the paler is the sand. Two of the beach stations, nos. 28 and 29, were situated in the Bay of Haifa. It can be seen from the figure that the sand from this area is among the palest on the coast. This characteristic implies relatively low iron oxide contents for these sands, so that they would make relatively pale or weakly coloured glasses. In addition, the calcium carbonate content of sand in the Bay of Haifa is around 15%, which would produce a soda-lime-silica glass with about 8.4% CaO. This is around the level required to produce a stable and workable glass. Thus, it appears that sand close to the mouth of the River Belus was the most suitable glassmaking sand on this part of the Mediterranean coast and was deliberately selected by glassmakers in the Greco-Roman period.

In fact, sands from many coastal locations in Israel contain lower $CaCO_3$ than those from the Bay of Haifa – sufficient to make a glass with, say, 5% CaO (Figure 4.3) – and it is therefore probable that other Levantine sands were used to make glass. For example, the late Byzantine/ early Islamic furnaces at Bet Eli'ezer, near Hadera, seem to have used sand with significantly lower CaO, but higher Al_2O_3, than those of Jalame and elsewhere and are likely to have used sand from a different location (Freestone, Gorin-Rosen, and Hughes 2000). Within the Roman period proper, lime and alumina contents of glasses vary within relatively narrow limits, but they can be used to subdivide the glass into meaningful compositional groups. It will be shown that the understanding provided by Pliny suggests that these were not all derived from Levantine sand.

The Volturno Sand

Having described 'the old method of producing glass' involving the Belus sand, Pliny goes on to describe the white sand that forms in the River Volturno and that is found on the beach between Cumae and Literno (about 50 km north of Naples). As discussed above, this sand was mixed with soda and fused to produce *hammonitrum*, which 'is again melted and forms pure glass, and is indeed a lump of clear colourless glass'.

The sand from the coast near the Volturno was inspected by Turner (1956b:Appendix, 300), who commented that 'nothing but grey-brown sand was seen throughout a several mile stretch of coast'. Mineralogical analysis revealed the presence of augite (an iron magnesium calcium silicate) in the sand, and Turner concluded that the sands had been modified since Pliny's day by the extensive volcanic activity in the region. More recently, Brill collected samples of beach sand from this coast and published their compositions (Brill 1999). Brill's analyses appear to corroborate the comments of Turner. The sands have relatively high aluminium and iron contents and are unsuitable for the production of weakly coloured or colourless glass. Further analyses of the Volturno sand have been published by Vallotto and Verità (2002) and closely agree with those of Brill. Although the sand of the Volturno is apparently most unsuitable for glassmaking, however, some characteristics are tantalisingly similar to those of certain types of Roman glass, as is shown below.

To approximate the composition of a hypothetical glass made from a specific sand such as that of the Volturno, the oxide concentrations in the sand may be reduced by 20%, to allow for the addition of 20% soda, Na_2O. Figure 4.4 shows the potash and alumina contents of such hypothetical Volturno glasses, based upon the sand compositions of Brill. They are compared with the compositions of typical Roman glass vessels from Italy, the glass of Aosta (*Augusta Praetoria*), analysed by Mirti and colleagues (1993). It is clear that the Volturno sands have potash and alumina contents that are far too high to allow the production of Roman glass. This suggests that the sands are too rich in minerals other than quartz (for example, feldspar: potassium aluminium silicate) that are likely to have been derived from volcanic activity in the region. The figure shows

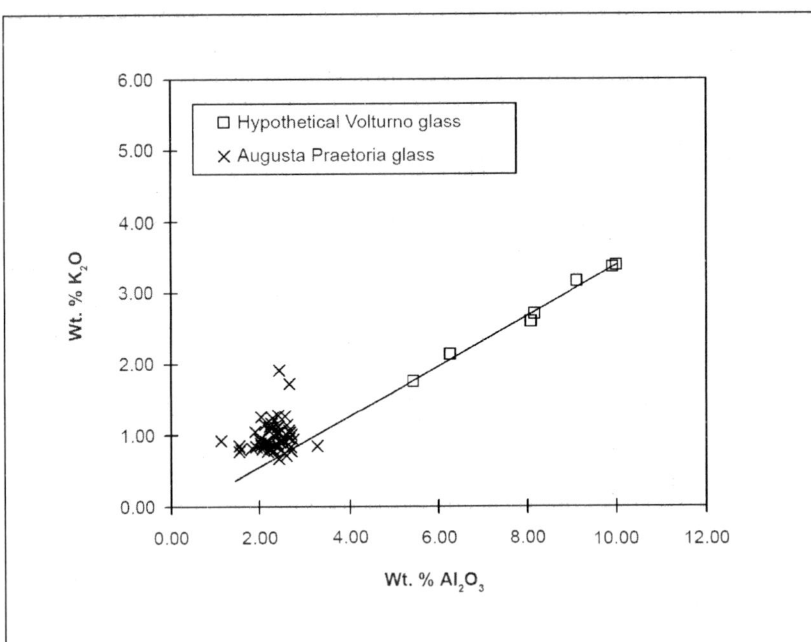

Figure 4.4 • Potash and alumina contents of hypothetical Volturno glasses, based on sands analysed by Brill (1999), compared with those of Roman glass from Augusta Praetoria (Mirti, Casoli, and Appolonia 1993). The trend line is a least squares fit to the Volturno glass compositions.

that there is a strong positive correlation between potash and alumina in the sands. The trend line through these points passes though the *Augusta Praetoria* glass group but is displaced away from the centre of it. Thus, it appears unlikely that the Volturno sands were used to produce 'typical' Roman glass and that even if the sand was refined or purified to reduce the total amount of the mineral impurities, it would not correspond to the Roman glasses.

Pliny is specific about the type of glass made from the Volturno sand; it is 'pure glass, and indeed a lump of clear colourless glass', according to the translation of Eicholz (1962, *Nat. Hist.* XXXVI:194). Recently, Roman transparent glasses that are *blue green* have been distinguished from those that are *colourless* (that is, the eye registers no significant tint in transmission) in a number of analytical studies (for instance, Heyworth, Baxter, and Cool 1990; Jackson 1994; Jackson et al. 1991). Figure 4.5 compares the compositions of a group of truly colourless Roman glass from Colchester, UK (Heyworth, Baxter, and Cool 1990), with the extension of the trend line in Figure 4.4. In contrast to the glass of mixed colours from Augusta Praetoria, the bulk of the colourless glasses group around the trend line, forming an elongate cluster with an axis parallel to, and

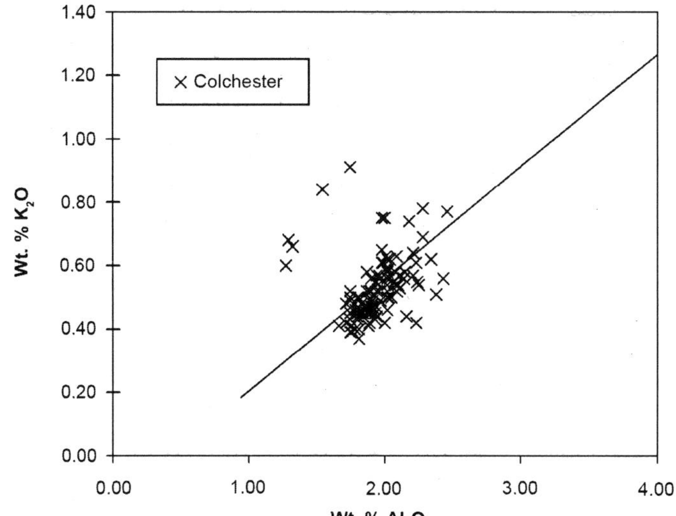

Figure 4.5 • Trend line from Figure 4.4, extrapolated down to compare with colourless glass from Roman Colchester (data from Heyworth, Baxter, and Cool 1990).

almost coincident with, the trend of the Volturno coastal sands. This coincidence of trend suggests that the compositions of the colourless glasses were controlled by a mineral assemblage that was closely similar to that responsible for the major variation in the Volturno sands. In the case of the colourless glasses, it appears that a sand from the coast in the region of the Volturno, but less contaminated by volcanic material than the samples collected and analysed by Brill and Vallotto and Verità, could have been used. It is not impossible that such a sand existed in Pliny's day but may have been worked out by the glassmakers or have been badly contaminated by later volcanic activity.

Pliny also comments that 'fossil (quarry) sand' was sometimes used in glassmaking (*Nat. Hist.* XXXVI:194) and also that the Volturno sand was taken 'wherever it is softest' (*Nat. Hist.* XXXVI:193). This could be read to imply that the Volturno sand required quarrying. Thus, although recent analyses of the Volturno sand suggest that it would not have been suitable for the manufacture of Roman glass, there is the possibility that a less contaminated sand, perhaps a geologically old sand that had to be excavated or quarried, was used to make colourless glass.

Additives

Shell

Virtually all analysed Roman glass has a soda-lime-silica base, containing between 4% and 10% CaO. Lime was an essential constituent: less than 4% would have rendered the glass very vulnerable to attack by moisture, whereas pure soda-silica glass is water soluble. There may have existed glasses containing exceptionally low concentrations of lime, but on the whole these will have disappeared as a result of 2,000 years of weathering.

The major sources of lime available to early glassmakers were the various naturally occurring forms of calcium carbonate, $CaCO_3$, principally limestone and shell. Where the source of the alkali in glass was the ash produced by burning plants or wood, then this, too, would have served as a source of lime (for instance, Freestone and Gorin-Rosen 1999). However, it is generally accepted that mineral soda was the alkali source used in Roman glass, so that lime is not likely to have entered glass via the use of ash. It has been considered by many authors that ancient glassmakers were not aware of lime as a specific additive (for example, Brill 1988; Turner 1956b) and that glasses were made essentially as mixtures of two materials, sand/silica and alkali. Thus, lime would have been added as a component of the sand, such as the lime-rich sands around the mouth of the River Belus, and the foregoing interpretation of Pliny's account is consistent with this view.

However, Pliny also comments that shell was added to glass in his day:

> Similarly, lustrous stones of many kinds came to be burnt with the melt and, then again, shells and quarry sand. (*Nat. Hist.* XXXVI:194)

To evaluate this statement, some compositional feature is required that discriminates between calcium carbonate in the form of shell and calcium carbonate in the form of limestone. Strontium, an element associated with calcium in the raw materials of glassmaking, offers this possibility. Shell imparts diagnostically high strontium contents to glass (c. 500 ppm) and diagnostic $^{87}Sr/^{86}Sr$ isotope ratios, which reflect those of Holocene sea water. Strontium and strontium isotope measurements on a number of Roman and Byzantine glasses suggest that the source of calcium in most of them is likely to have been shell (Freestone et al. 2003; Wedepohl and Baumann 2000); large numbers of analyses carried out on Roman glass in the past few years (see works cited by Heyworth et al., Mirti et al., and Jackson et al.) reveal elemental strontium in the order of 500 ppm, again suggesting that the bulk of the lime was indeed present in the batch as shell. On initial consideration, this appears to support Pliny's statement that shell was added to the glass batch. However, these results are also consistent with the use of beach sand, which contained abundant shell fragments – for example, the Belus and other Levantine sands – and do not necessarily imply the deliberate addition of shell as a separate component.

If shell was intentionally added to glass, then it is unlikely to have been when the Belus sand was used for, as has been seen, all Levantine coastal sand contains sufficient calcium carbonate to yield a stable glass. Furthermore, in that the Belus sand was characteristic of the 'old way' of making glass, it seems likely that if shell was used, then it is likely to have been in the 'new' approach that used the Volturno sand. In addition, given that he lived in the region, Pliny is likely to have been more informed about the Volturno glassmaking method than any other.

To sum up, the glassmaking sand with which we are most familiar today, the Syro-Palestinian coastal sand near the mouth of the River Belus, contained sufficient lime to render any further addition unnecessary. The coastal sand in the vicinity of the River Volturno is described as a raw material for colourless glass by Pliny, but modern analyses have not yet detected here sand sufficiently low in alumina, potash, and iron to match Roman glass. However, certain compositional traits may be tentatively interpreted to suggest a relationship between the Volturno sands and Roman colourless glass, implying that a sand of the appropriate composition may have existed in the region. If so, it might have been necessary to add lime to the glass, and this is likely to have been in the form of shell, as described by Pliny.

Lustrous Stones

Pliny notes that 'lustrous stones of many kinds came to be burnt with the melt' (*Nat. Hist.* XXXVI:194). Healy (1999) has recorded a wide range of lustrous stones known to Pliny, which may have been beneficial in glassmaking or colouration, including opal, agate, and onyx; sapphire and ruby; malachite and azurite. However, these were all of value for decorative purposes, and it seems unlikely that they would have been destroyed to make glass, which was essentially an imitation stone. Furthermore, it is unlikely that the copper carbonates, azurite and malachite, were used to colour glass blue. Traces of tin are associated with copper blue vitreous materials from the Bronze Age onward and indicate that the scale produced by heating scrap copper alloy in air was the main source of copper in glassmaking (Sayre and Smith 1967).

A lustrous mineral that was extensively used in Roman glass technology but that apparently goes unmentioned by Pliny in this context was stibnite, antimony sulphide, Sb_2S_3. Not only were most opaque colours dependent on the development of antimony compounds (Mass et al. 1998), but antimony was also added as a decolouriser (oxidant) in the manufacture of colourless glass (Sayre 1963). Stibnite is by far the most common ore of antimony; it has a metallic lustre, brilliant on fresh surfaces (Mason and Berry 1968), and its extensive use in Roman glassmaking suggests that it would have deserved mention in Pliny's discussions with his sources of information. We are unlikely ever to be sure of the meaning of 'lustrous stones', but stibnite would appear to be more likely than many others that have been mentioned in this context.

Magnes Lapis

Pliny mentions a number of other additives to glass, including *magnes lapis*, the 'magnet stone'. This mention has given rise to a great deal of discussion. It has been considered that the 'magnet stone' is unlikely to have been magnetite (Healy 1999), since this would not have been beneficial to the production of colourless glass. The idea that Pliny may have been referring to manganese has been favoured by some authors, as summarised by Turner (1956a), who did not commit himself to a particular interpretation. In more recent interpretations, 'magnesian limestone' tends to be preferred (Healy 1999:353, footnote). Although it is unclear that we will ever be sure just what was being referred to as *magnes lapis*, it is possible to place some constraints on this term.

In Pliny's own description of *magnes lapis*, earlier in the book (*Nat. Hist.* XXXVI:126–29), he derives his description from that of Sotacus, the author of a work on stones who wrote some three centuries earlier. According to Pliny's summary, Sotacus described five kinds of naturally occurring magnet. However, he notes that that 'the most worthless kind is that of Magnesia in Asia Minor, which is white, has no power of attracting iron, and resembles pumice', those from the Troad were black, but female – that is, had no magnetic force – and so on. It appears that there was no fixed concept as to the physical properties of *magnes lapis* – even the ability to attract iron was not a requirement – and that this term could encompass a wide range of materials.

The possibility that magnetite may have been added to glass cannot be dismissed. Clearly, the addition of iron oxide is undesirable if it is wished to produce a colourless glass, but magnetite may have been a desirable additive in the production of some colours. For example, some Roman glasses coloured opaque red by copper contain elevated iron oxide, which may have acted as a reducing agent to promote the precipitation of the copper colourant phase (Freestone, Stapleton, and Rigby 2003). Magnetite may have been added as the active ingredient for this purpose, although it would have been less effective as a reductant than the scale produced by heating scrap iron. Iron is also elevated in many Roman blue glasses and is associated with the cobalt colourant. Magnetite can be a major component of cobalt ores; for example, it occurs in the famous cobalt deposits of Kashan, Iran, which were exploited in the medieval Islamic world (Dayton and Bowles 1977). It might well have appeared to the glass-maker that the material that they were adding to colour their glass blue was indeed a type of magnetite.

The idea that magnesian limestone (a rock composed of calcium and magnesium carbonates), or even magnesite (magnesium carbonate), was added to glass has been considered by some authors (Healy 1999; Knoll et al. 1979) but is unlikely, in view of the magnesium oxide contents of most Roman glass, which are very low, at around 0.5%, and may be explained by the MgO present in the glassmaking sand (for instance, Brill 1988). Furthermore, it has been observed above that the lime in Roman glass is likely to have been derived mainly from shell, rather than limestone.

Notwithstanding the possible addition of magnetite to some coloured glasses, on the basis of its ubiquity in Roman glass, the best candidate for *magnes lapis* would appear to be manganese oxide. Black pyrolusite (MnO_2) could readily have been included as one of the nonmagnetic, or female, varieties of *magnes lapis* in the classification of Sotacus. From the 2nd century B.C.E., manganese becomes a frequent component in colourless, or weakly coloured, glass (Sayre 1963). By the time that Pliny wrote, it was a standard additive to most glass, probably to minimise the colour due to iron. It would surely have been worthy of comment in his account of glassmaking.

Evidence for Changes in the Location of Production

When Pliny comments that the manufacture of glass using the Belus sand was 'the old way' and that the modern way was to use the sand of the coast near the River Volturno, he also notes that 'sand is similarly blended in the Gallic and Spanish provinces'. Here he is apparently indicating that glass was made from raw materials in France and Spain. These various glassmaking locations should be reflected in the chemical compositions of glass from the period, since each is likely to have meant the use of a different sand source. Although the composition of Roman glass from some regions is not well known (for example, there appear to be few chemical analyses of Roman glass from Spain), sufficient data have been compiled to demonstrate that all Roman glass is very similar in composition. However, within this apparently homogeneous assemblage, some trends and subgroupings are becoming apparent.

Figure 4.6 compares the alumina and lime contents of some Roman glasses of the 1st to 3rd centuries. These oxides are potentially useful because, in at least some glasses, both are likely to have been almost entirely imparted to the glass by the glassmaking sand. Groups that are distinguishable on the basis of these oxides are therefore likely to represent glasses made from different sands.

The three glass groups represented in Figure 4.6 are (1) colourless glass of the 1st–3rd centuries C.E. from Britain (Heyworth et al. 1990; Mortimer and Baxter 1996), (2) weakly coloured blue-green glass of the 1st–3rd centuries from Britain (Jackson 1994; Jackson et al. 1991), and (3) polychrome mosaic glass vessels and cameo glasses of the 1st century B.C.E. – 1st century C.E. (Bimson and Freestone 1983; Stapleton 2003). The glasses compositions have been *reduced* so that colourant and decolourant elements have been removed and the remaining oxides recast to 100% (see Brill 1999).

The early polychrome mosaic glass vessels were produced between 25 B.C.E. and 25 C.E., up to a century or so before Pliny wrote the *Natural History*. They are therefore likely to represent glasses made by 'the old method', that is, using the sand from the mouth of the River Belus. Pliny's 'new' method of making glass involved the production of glass that was colourless. The archaeology of glass vessels is consistent with this, since it suggests that colourless glass became common in the third quarter of the 1st century C.E. (Price and Cottam 1998:15–16), at about the time

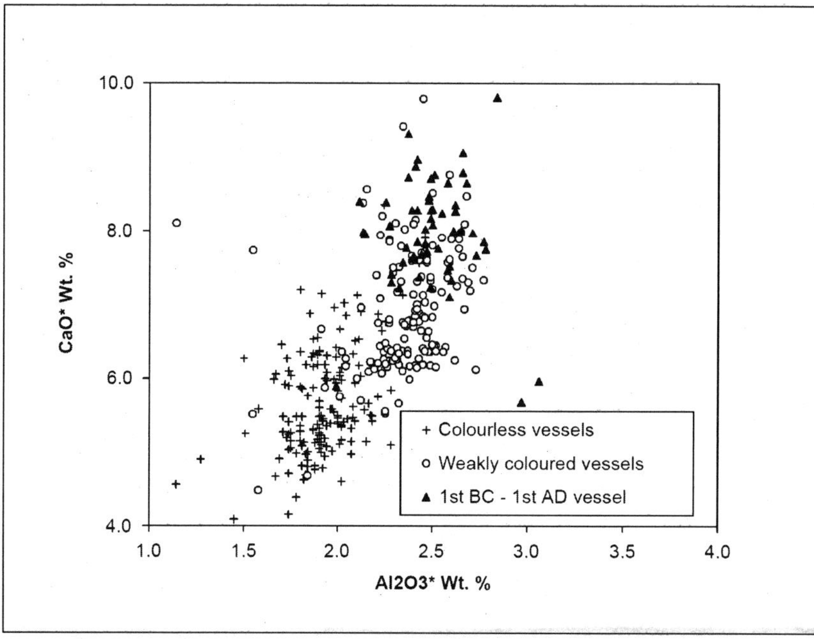

Figure 4.6 • Lime and alumina concentrations in various Roman glass groups; asterisks indicate that the compositions have been reduced (that is, colourants removed and bulk recast to 100%; see text).

that Pliny was writing. Figure 4.6 shows a strong separation between the early polychrome glass and the later colourless glass, confirming a major change in sand source.

The weakly coloured glass of the 1st to 3rd centuries occupies the region between the other groups and, in particular, shows some overlap with the early group (Figure 4.6). It is widely considered that these 'typical' Roman glasses were also made from Belus type sand (Nenna et al. 1997; Vallotto and Verità 2002). If this is so, then some compositional drift in sand composition must have occurred during the 1st century C.E., to the lower lime contents shown by these glasses. Possibly sand was obtained from a slightly different location on the coast of Syria-Palestine (Figure 4.3). However, it is probably prudent at this stage to keep an open mind to the possibility that some of the glass was made from sand 'blended in the Gallic and Spanish provinces'.

This brief examination of just the major elements of Roman glass, conventionally considered a very homogeneous material, has already revealed meaningful groups that are consistent with Pliny's account. It is possible that further groupings will be revealed by more detailed analyses.

Discussion: Pliny and the Archaeometry of Glass Production

A synthesis of textual and archaeological evidence may be anticipated to enlighten in a number of ways, and several of these are illustrated by the present consideration of Pliny on glassmaking.

An important aspect of textual information is that it may point to interpretations of archaeological data that would not have been obvious, and might have taken much longer to attain, in the absence of the written word. Perhaps most important are Pliny's remarks on Egyptian soda sources, which have underpinned our understanding of Roman glass composition for many years. A further example is the significance of the Belus River source of sand. In spite of a number of excavations of Byzantine and Early Islamic glassmaking furnaces in the Levantine region, the furnaces that putatively made the early Roman glass remain to be discovered. Furthermore, the glasses from these later furnaces differ in composition from the early Roman glasses. However, the abundant evidence for later glassmaking suggests a strong regional tradition and together with the comments of Pliny, Strabo, and others strongly supports the view that glass was made there in earlier Roman times.

Given the composition of modern sands around the mouth of the River Volturno, it must be considered unlikely that this location would ever have been considered as a source of good glassmaking sand on the basis of archaeology alone, unless a group of furnaces were discovered. However, Pliny's comments have prompted comparisons of the compositional data for the Volturno sands and those for Roman glass. Such comparisons have allowed some rather tenuous similarities in compositional trends to be drawn out, which might just indicate a relationship between the sands of the region and the glass. At the very least, a possible Volturno source would appear to be worthy of further investigation.

There are other areas, however, where the reading of Pliny in the absence of suitable archaeological and scientific evidence has proved far from helpful and, indeed, has been misleading. The comments of Turner (1956b) that Pliny's account of glassmaking was 'obscure' have already been noted, and Healy, too, considers it to be 'confused' (1999: 354). Furthermore, this apparent obscurity allows a range of potential interpretations to be read into the text. Thus Turner (1956b) suggests that Roman glassmaking involved a 'low-temperature' fritting process followed by a conversion to glass in crucibles, and Healy (1999) also draws parallels with modern practice, suggesting an initial low temperature firing. It has been seen that current archaeological and chemical analytical evidence suggest an initial *high temperature* fusion direct to a primary glass, which did not involve the use of crucibles. In this case, the archaeology has allowed us to read Pliny probably as he intended. Here, a comparison of archaeology and text has demonstrated that an area of the text that appeared 'confused' was more or less correct, although it was not expressed with the precision and close attention to clarity that a modern technical writer would strive to attain. It must be remembered that Pliny was not trying to write a manual or recipe book but an encyclopaedia; the precise step-by-step details of the glassmaking process are likely to have been less important than the general concepts. In other areas of ambiguity, the scientific analysis of the glass provides the data against which each of the various interpretations possible may be tested, as in the case of 'magnes lapis', 'lustrous stones', and the addition of shell to glass, discussed above.

The extent to which any two lines of evidence are complementary depends critically on their quality, and the detail and clarity of both the written word and the archaeological record are notoriously variable. The present exercise demonstrates that in order to understand Roman glass production with confidence, both are needed. The text provides a number of key insights that might have taken a great deal of archaeological work to elicit. However, archaeology was able to clarify some apparently obscure areas and constrain the interpretation of others.

There are, of course, aspects of life where textual evidence is potentially far more powerful than archaeology. Ideology and belief are areas that are in general far more easily conveyed through words than through the analysis of objects or structures. However, technology is an area where textual evidence may be quite limited, not least because technical processes are frequently of little interest to the literate elites of complex societies. Even where a description of a process or material is given, the absence of a systematic classification of materials will always have resulted in a degree of ambiguity in early texts. Furthermore, the details of an operational sequence are extremely difficult to commit to text, as anyone who has attempted to follow the instructions included with a piece of modern, flat-packed furniture will testify. Archaeological science offers powerful tools to interpret such processes and should be considered the touchstone against which their accuracy is evaluated, rather than depending on parallels with modern industrial practices and understanding, as it has often been necessary to do in the past.

Acknowledgments

I am very grateful to M. Martinón-Torres and Th. Rehren for organising the meeting on Archaeology and Texts and for prompting me to write this paper. The manuscript was kindly commented on by S. Bowman, J. Price, an anonymous referee, and the editors. However, I am fully responsible for any errors or omissions.

References

Aldsworth, F., G. Haggarty, S. Jennings, and D. Whitehouse. 2002. 'Medieval glass-making at Tyre, Lebanon', *Journal of Glass Studies* 44:49–66.

Bimson, M., and I. C. Freestone. 1983. 'An analytical study of the relationship between the Portland Vase and other Roman cameo glasses', *Journal of Glass Studies* 25:55–64.

Brill, R. H. 1988. 'Scientific investigations', in G. D. Weinberg (ed.), *Excavations at Jalame: Site of a Glass Factory in Late Roman Palestine.* Columbia: University of Missouri, Columbia, 257–94.

———. 1999. *Chemical Analyses of Early Glasses* Corning, NY: Corning Museum of Glass.

Brill, R. H., and J. F. Wosinski. 1965. 'A huge slab of glass in the ancient necropolis of Beth She'arim', *Proceedings of the 7th International Congress on Glass.* Brussels, Paper 219, 4–11.

Cable, M. 2002. *The World's Most Famous Book on Glassmaking: The Art of Glass by Antonio Neri* (C. Merrett, trans.). Sheffield: Society of Glass Technology.

Dayton, J. E., and J. Bowles. 1977. 'Abul Qasim of Kashan, and the problem of Persian glazing', *Annali dell'Istituto di Napoli* 37:143–52.

Eicholz, D. E. 1962. *Pliny – Natural History Books 36–37.* Cambridge, MA: Harvard, Loeb Classical Library.

Emery, K. O., and D. Neev. 1960. 'Mediterranean Beaches of Israel', *Geological Survey of Israel Bulletin* 26:1–23.

Foy, D., M. Picon, and M. Vichy. 2000. 'Les Matières premières du verre et la question des produits semi-finis, Antiquité et Moyen Âge', in *Arts du Feu et Productions Artisanales*: XXe Recontres Internationales d'Archéologie et d'Histoire d'Antibes. Antibes: Éditions APDCA, 419–32.

Foy, D., M. Vichy, and M. Picon. 2000. 'Lingots de verre en Méditerrané occidentale', in *Annales du 14th Congrès de l'Association pour l'Histoire du Verre*, 51–57.

Freestone, I. C., and Y. Gorin-Rosen. 1999. 'The great glass slab at Bet Shearim: An early Islamic glass-making experiment?', *Journal of Glass Studies* 41:105–16.

Freestone, I. C., Y. Gorin-Rosen, and M. J. Hughes. 2000. 'Primary glass from Israel and the production of glass in late antiquity and the early Islamic period', in M. D. Nenna (ed.), *La Route du Verre: Ateliers primaires et secondaires de verriers du second millénaire av. J.-C. au Moyen-Âg.* Lyon: Travaux de la Maison de l'Orient *Mediterranée* 33, 65–83.

Freestone, I. C., R. Greenwood, and Y. Gorin-Rosen. 2002. 'Byzantine and early Islamic glassmaking in the Eastern Mediterranean: Production and distribution of primary glass', in G. Kordas (ed.), *Hyalos – Vitrum – Glass.* Athens: Glasnet Publications, 167–74.

Freestone, I. C., M. Ponting, and M. J. Hughes. 2002. 'Origins of Byzantine glass from Maroni Petrera, Cyprus', *Archaeometry* 44:257–72.

Freestone, I. C., K. A. Leslie, M. Thirlwall, and Y. Gorin-Rosen. 2003. 'Strontium isotopes in the investigation of early glass production: Byzantine and early Islamic glass from the Near East', *Archaeometry* 45:19–32.

Freestone, I. C., C. P. Stapleton, and V. Rigby. 2003. 'The production of red glass and enamel in the Later Iron Age, Roman and Byzantine periods', in C. Entwistle (ed.), *Through a Glass Brightly: Studies in Byzantine Art and Archaeology Presented to David Buckton.* Oxford: Oxbow, 142–54.

Goldsmith, V., and A. Golik. 1980. 'Sediment transport model of the southeastern Mediterranean coast', *Marine Geology* 37:147–75.

Gorin-Rosen, Y. 1995. 'Hadera, Bet Eli'ezer', *Excavations and Surveys in Israel* 13:42–43.

———. 2000. 'The ancient glass industry in Israel: Summary of new finds and new discoveries'. In M. D. Nenna (ed.), *La Route du Verre: Ateliers primaires et secondaires de verriers du second millénaire av. J.-C. au Moyen-Âge*. Lyon: Travaux de la Maison de l'Orient Mediterranée 33, 49–64.

Healy, J. F. 1981. 'Pliny the Elder and ancient mineralogy'. *Interdisciplinary Science Reviews* 6:166–80.

———. 1991. *Pliny the Elder: Natural History – A Selection*. Harmondsworth: Penguin.

———. 1999. *Pliny the Elder on Science and Technology*. Oxford: Oxford University Press.

Henderson, J. 1985. 'The raw materials of early glass production', *Oxford Journal of Archaeology* 4:267–91.

Heyworth, M. P., M. J. Baxter, and H. Cool. 1990. *Compositional analysis of Roman glass from Colchester, Essex*. Ancient Monuments Laboratory Report 53/90, English Heritage.

Ignatiadou, D., E. Dotsika, A. Kouras, and Y. Maniatis. 2005. 'Nitrum Chalestricum: The natron of Macedonia', in H. Cool (ed.), *Annales du 16e Congres de l'Association Internationale pour l'Histoire du Verre*, 64–67.

Jackson, C. M. 1994. Appendix A1, in M. J. Baxter, *Exploratory Multivariate Data Analysis in Archaeology*. Edinburgh: Edinburgh University Press, 228–31.

Jackson, C. M., J. R. Hunter, S. E. Warren, and H. E. M. Cool. 1991. 'The analysis of blue-green glass and glassy waste from two Romano-British glass-working sites', in E. Pernicka and G. A. Wagner (eds.), *Archaeometry '90*. Basel: Birkhäuser Verlag, 295–305.

Jones, W. H. S. 1963. *Pliny: Natural History, Books XXVIII–XXXII*. Cambridge, MA: Harvard University Press.

Knoll, H., A. Locher, R. C. A. Rottländer, O. Schaaber, H. Scholze, G. Schulze, G. Strunk-Lichtenberg, and D. Ullrich D. 1979. 'Glasherstellung bei Plinius dem Älteren', *Glastechnische Berichte* 52:265–70.

Kock, J., and T. Sode. 1994. *Glass, Glass Beads and Glassmakers in Northern India*. Vanlose, Denmark: THOT.

Lucas, A., and J. R. Harris. 1962. *Ancient Egyptian Materials and Industries* (4th ed.). London: Edward Arnold.

Mason, B., and L. G. Berry. 1968. *Elements of Mineralogy*. San Francisco: W. H. Freeman.

Mass, J. L., R. E. Stone, and M. T. Wypyski. 1998. 'The mineralogical and metallurgical origins of Roman opaque coloured glasses', in P. McCray (ed.), *The Prehistory and History of Glassmaking Technology* (Vol. 3). Ceramics and Civilization 3. Westerville, OH: American Ceramics Society, 121–44.

Mirti, P., A. Casoli, and L. Appolonia. 1993. 'Scientific analysis of Roman glass from Augusta Praetoria', *Archaeometry* 35:225–40.

Mortimer, C., and M. J. Baxter. 1996. 'Analysis of samples of colourless Roman vessel glass from Lincoln', *Ancient Monuments Laboratory Report* 44/96, English Heritage.

Nenna, M-D., M. Vichy, and M. Picon. 1997. 'L'Atelier de verrier de Lyon, du Ier siècle après J.-C., et l'origine des verres "Romains"', *Revue d'Archèomètrie* 21:81–87.

Nenna, M-D., M. Picon, and M. Vichy. 2000. 'Ateliers primaire et secondaires en Égypt à l'époque gréco-romaine', in M. D. Nenna (ed.), *La Route du Verre:*

Ateliers primaires et secondaires de verriers du second millénaire av. J.-C. au Moyen-Âge. Lyon: Travaux de la Maison de l'Orient Mediterranée 33, 97–112.

Price, J., and H. E. M. Cool. 1991. 'The evidence for the production of glass in Roman Britain', in D. Foy and G. Sennequier, *Ateliers de Verriers de l'Antiquité à la periode pré-industrielle*. Rouen: Association Française pour l'Archéologie du Verre, 23–30.

Price, J., and S. Cottam. 1998. *Romano-British Glass Vessels: A Handbook*. York: Council for British Archaeology.

Rehren, Th., and E. B. Pusch. 2005. 'Late Bronze Age glass production at Qantir-Pirimesses, Egypt', *Science* 308:1756–58.

Sayre, E. V. 1963. 'The intentional use of antimony and manganese in ancient glasses', in F. R. Matson and G. E. Rindone (eds.), *Advances in Glass Technology, Part 2*. New York: Plenum Press, 263–82.

Sayre, E. V., and R. W. Smith. 1967. 'Some materials of glass manufacturing in Antiquity', in M. Levey (ed.), *Archaeological Chemistry: A Symposium*. Philadelphia: University of Pennsylvania Press, 279–311.

Schreurs, J. W. H., and R. H. Brill. 1984. 'Iron and Sulfur related colours in ancient glasses', *Archaeometry* 26:199–209.

Shortland, A. J. 2000. *Vitreous Materials at Amarna: The Production of Glass and Faience in 18th Dynasty Egypt*. BAR International Series, S827. Oxford: Archaeopress.

———. 2004. 'Evaporites of the Wadi Natrun: Seasonal and annual variation and its implication for ancient exploitation', *Archaeometry* 46:497–516.

Sode, T., and J. Kock. 2001. 'Traditional raw glass production in northern India: The final stage of an ancient technology', *Journal of Glass Studies* 43:155–69.

Stanley, D. J., Y. Mart, and Y. Nir. 1997. 'Clay mineral distributions to interpret Nile cell provenance and dispersal: II. Coastal plain from Nile delta to northern Israel', *Journal of Coastal Research* 13:506–33.

Stapleton, C. P. 2003. 'Composition and technology of mosaic vessel glass from the early Roman Empire', *Annales du 15e Congrès de l'Association Internationale pour l'Histoire du Verre*, 29–32.

Tal, O., R. E. Jackson-Tal, and I. C. Freestone. 2004. 'New evidence of the production of raw glass at Late Byzantine Apollonia-Arsuf (Israel)', *Journal of Glass Studies* 46:51–66.

Thompson, D. (ed.). 1995. *The Concise Oxford Dictionary*. Oxford: Clarendon Press.

Turner, W. E. S. 1956a. 'Studies in ancient glasses and glassmaking processes: Part III, The chronology of the glassmaking constituents', *Journal of the Society of Glass Technology* 40:39–52.

———. 1956b. 'Studies in ancient glasses and glassmaking processes: Part V, Raw materials and melting processes', *Journal of the Society of Glass Technology* 40: 277–300.

Vallotto, M., and M. Verità. 2002. 'Glasses from Pompeii and Herculaneum and the sands of the rivers Belus and Volturno', in J. Renn and G. Castagnetti (eds.), *Homo Faber: Studies on Nature, Technology and Science at the Time of Pompeii*. Rome: 'L'Erma' di Bretschneider, 63–73.

Wedepohl, K. H., and A. Baumann. 2000. 'The use of marine molluskan shells for Roman glass and local raw glass production in the Eifel area (Western Germany)', *Naturwissenschaften* 87:129–32.

CHAPTER 5

Ptolemaic and Roman Memphis as a Production Centre

SALLY-ANN ASHTON

ABSTRACT *Alexandria is traditionally viewed as an artistic as well as intellectual centre in Ptolemaic and Roman Egypt; there is, however, little archaeological evidence to support this. The role played by Memphis during these two periods is often underestimated, although the archaeological record suggests that both the sacred and the domestic areas of the city were of considerable importance, particularly during the Ptolemaic period, when it enjoyed two periods of royal residency under Ptolemy I and Ptolemy V. This chapter considers the excavated material from Flinders Petrie's seasons at Memphis, currently held in the Petrie Museum of Egyptian Archaeology at University College London, which includes terracotta and faience objects, a 1st-century* C.E. *faience factory, and plaster moulds associated with both the production of forms of representation and the manufacturing of bronze. The question of the role that was played by Memphis in the production of Ptolemaic and Roman material culture is reassessed, as well as the close association between stone and bronze sculptural and terracotta representations; furthermore, ceramic and faience vessels in the Greek, Roman, and Egyptian styles are explored. The results of a survey of the archaeological evidence suggests that, in the Ptolemaic and early Roman periods, Memphis was an active production centre with links between the forms of artistic media to the religious activities in the city, thus compelling a reassessment of the accepted knowledge and offering a more accurate representation of Egypt in the Ptolemaic and early Roman Imperial periods.*

Introduction

Alexandria is traditionally viewed as an artistic as well as a cultural centre in Ptolemaic and Roman Egypt (see Stewart 1996:231–46 for history); there is, however, little archaeological evidence to support this hypothesis. These periods run from 323–30 B.C.E. (Ptolemaic) and the first two centuries of Roman occupation from 30 B.C.E. to around 200 C.E. The term 'Alexandrian' is generally used to describe Classical, that is to say, Greek or Roman, material culture or, in some cases, simply a style. The reason for this interpretation is that the first scholars to work seriously on the Ptolemaic and Roman periods in Egypt – for example, Achille Adriani and Giuseppe Botti – were by training Classicists, and they naturally gravitated toward the city that had produced literary and historical works under the patronage of the early Ptolemaic rulers in the Mouseion. The role played by Memphis during these two periods is often underestimated, although the archaeological record suggests that both the sacred and the domestic areas of the city were of considerable importance during the Ptolemaic period, when it enjoyed two periods of royal residency under Ptolemy I and V, and a continued close connection with the royal house. It is only recently that the same group of scholars has paid attention to Memphis, Egypt's second city, during these two periods.

One scholar in particular has provided a detailed and compelling account of Memphis in the Ptolemaic period. D. J. Thompson's *Memphis under the Ptolemies* has illustrated that Memphis was an important cultural as well as religious centre. Thompson predominantly used textual evidence to obtain a greater understanding of the city and its inhabitants, most notably those involved in religion, from the Apis bulls that were housed at the site, to the Sarapieion twins who complained bitterly about their treatment as recluses in the sanctuary of the god (Thompson 1988:155–65). Although Thompson refers briefly to industries at the site, the finds from Sir Flinders Petrie's excavations at Memphis, now in the Petrie Museum of Egyptian Archaeology (University College London), have never fully been explored. Petrie was the first archaeologist to realise the potential of Memphis, and, unlike many Egyptologists, he was interested in the Ptolemaic and Roman material in spite of its later date. As an excavator, Petrie was interested in technology and used many of his finds from Memphis to illustrate manufacturing techniques. However, excavations undertaken in the early 1900s do not offer the same stratified evidence that, one would hope, modern archaeologists are able to provide. The result of developments in archaeological techniques and recording mean that 21st-century scholars might easily throw up their hands in dismay, and dismiss this important material. Nevertheless, the objects collected by Petrie offer a unique insight into the developments and changes of technologies, fashions, and the interactions among Greek, Roman, and Egyptian artists during the Ptolemaic and Roman periods.

That Memphis provided a suitable environment for such innovation had been noted as early as the 1st century B.C.E., when Strabo, a Greek visitor to the site, wrote: 'The city is large and populous, ranking second after Alexandria, and consisting of mixed races of people, like those who have settled together at Alexandria' (Strabo, *Geography* 17.1.30–32). The material from Memphis discussed in this chapter can be divided into four areas of technology: (1) faience vessels; (2) stone sculpture; and (3/4) bronze and terracotta figurine production. Each group has its problems, and in addition to illustrating these categories of finds I also attempt to identify questions relating to the evidence from Petrie's excavations.

Faience

Egyptian faience is a glazed silica-rich material, composed of crushed quartz or sand with small amounts of lime, and a soda-based glaze. Its main ingredient was quartz, obtained from calcareous sand or crushed pebbles, mixed with mineral natron or plant ash as alkali source for the glaze, and a small amount of ground copper minerals as colourant. After the paste was modelled into the desired shape, it was fired in a kiln. There were different ways to achieve the glazed surface: by efflorescing the alkali salts on the surface through evaporation of the water from the brine mixed with the body, by applying the glazing mixture externally, or by cementation, in which the modelled object was placed in an alkali-rich medium before it was fired. Upon firing, the lime and soda fluxed and cemented the silica-rich body, and a soda-lime-silica glaze formed on the surface, typically coloured blue by the copper pigment.

In Memphis, evidence of faience production was found at two sites: Kom el-Qalama, which Petrie believed to be Ptolemaic in date, and Kom Helul, which he dated to the Roman period (Petrie 1909:3). It seems that Petrie's only reason for dating the Kom el-Qalama site to the Ptolemaic period was the size of stacking cones that he found there (Petrie 1911) (Figure 5.1), although it should be noted that there are several examples of Ptolemaic vessels believed to have come from Memphis, which are now housed in the Petrie Museum (Figure 5.2). This material could per-haps be associated with the earlier site but is not marked as such, which illustrates the need not only for marking material on site but also for attention to the processing and cataloguing of finds. Such lost information is a constant cause of frustration when working on the Memphite material, and it exemplifies the need for a reasonable knowledge of the finds, in order to link them with Petrie's reports. It is also worth noting that there is no industrial waste dating to the Ptolemaic period,

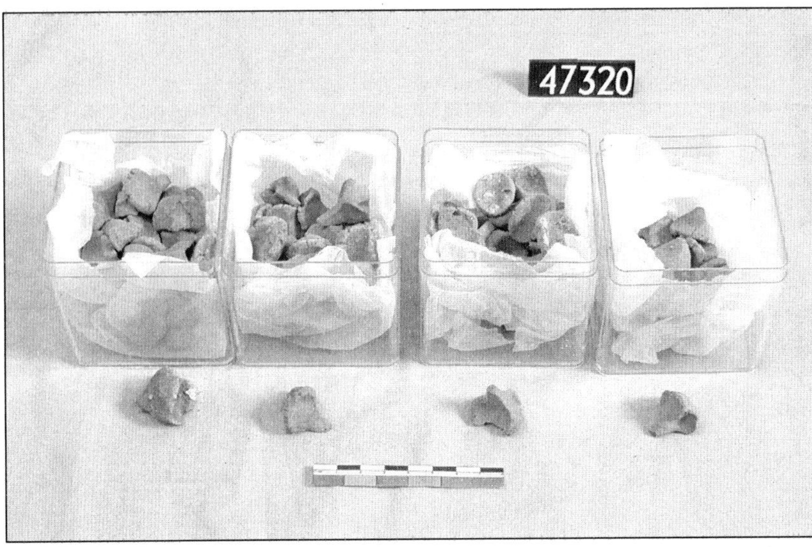

Figure 5.1 • Various sizes of ceramic cones used as supports for vessels in the kiln during firing (UC 47320) (© Petrie Museum of Egyptian Archaeology, University College London).

Figure 5.2 • Rim fragment of light green faience vessel, raised leaf design on exterior with raised blue dot at the apex of the complete centre leaf (UC 2339) (© Petrie Museum of Egyptian Archaeology, University College London).

and the fragments of bowls from Memphis, as illustrated in Ashton (2003b:42–43 and 53–58), may simply have been votives from one of the many sacred areas at the site.

Kom Helul was recorded in more detail than Kom el-Qalama and has remained the focus of work on faience production at Memphis. Petrie found and excavated a kiln there (1909:14–15, Pl. XLIX), but in his report he states that the industrial waste had probably not been produced at Kom Helul but had been moved there from neighbouring kilns and dumped.

More recent excavations confirm Petrie's observations, with pottery dating from at least the early Ptolemaic to early Roman periods in disturbed layers.[1] The faience fragments from the site are often in the form of kiln wasters and so illustrate that in the early Roman period, somewhere in Memphis there was a production centre for faience vessels. The finds have not only allowed archaeologists to record the date of production but have also prompted attempts to identify the manufacturing process of Roman faience (Petrie 1911 and Nicholson 1998). However, caution is necessary. Both excavators at Kom Helul have recognised that the material had been dumped at the site, but this has not prevented them from attempting to piece together all the material in order to obtain a greater understanding of manufacturing processes. The problem here is that not all the associated finds may have been concerned with the manufacture of faience, and yet people working on the material have often failed to separate that concerned with faience from that concerned with glass. One particular problem is understanding how the frit and frit-lined vessels that were excavated at the site were used (Figure 5.3). It has recently been suggested that such material was used for the manufacture of Egyptian blue (Th. Rehren and M. S. Tite, pers. comm. 2005).

Although the problems with the actual manufacturing processes may continue to be debated, and it may be necessary to recognise and accept that the Memphis material may never solve them, the very fact that faience vessels continued to be manufactured in the early Roman period at Memphis is of considerable interest. Furthermore, the recipe used for faience during this period is different from that of Ptolemaic Egyptian and Greek-style faience, resulting in a visual difference of the glaze and core (Ashton 2003b:39–58). The Memphis material illustrates that the differences in the matrix and glaze of Ptolemaic and Roman faience are chronological rather than geographical, if we assume that Petrie did find earlier kilns at the site. The evidence from Kom Helul suggests mass production of vessels, probably for use as votives at the temples in the city and at Saqqara. Vessels were stacked in kilns and fired; the glazes were applied sloppily, often resulting in residue accumulating in the bottom of bowls, forming a thick blue pond, unnoticed by Petrie and

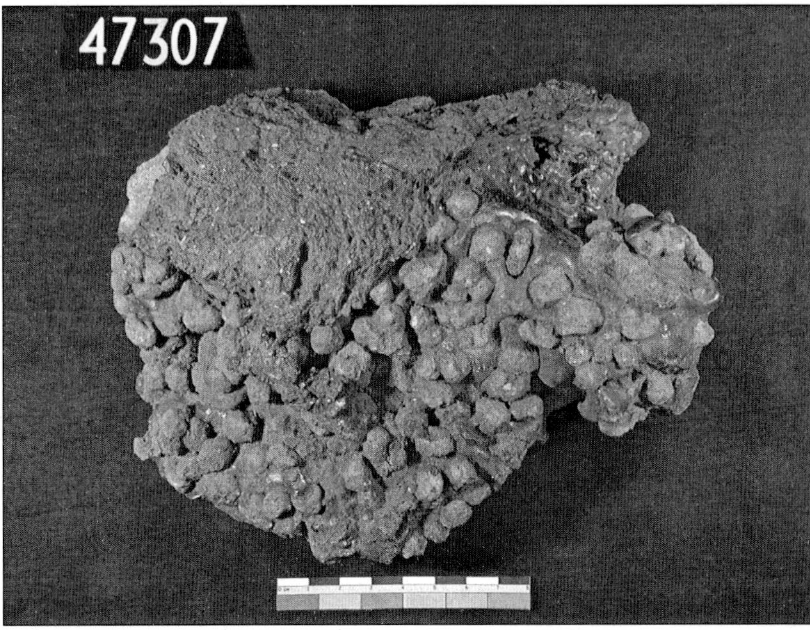

Figure 5.3 • Body fragment of an industrial pottery vessel in the form of a jar with balls of blue frit on both sides and a further five fragments of larger pieces of frit; there is also a large piece of red baked clay attached to one side of the fragment (UC 47307) (© Petrie Museum of Egyptian Archaeology, University College London).

later by Nicholson, who suggested that vessels were fired upside down (Nicholson 1998:255) (Figure 5.4). The extent of information that can be extracted from the faience finds is determined by the amount of knowledge held by an individual scholar. For those who are able to date the vessel forms (Ashton 2003b:51–58), the material reveals a process of change in manufacture, which mirrors that of Egypt's political climate, not only in the forms of vessels but also in their composition and the methods of manufacture. Within two generations of Roman occupation, the production of material culture has moved from a centrally controlled and patronised system to a more localised and less supported structure (Ashton 2003b:29–37). Without the archaeological evidence, there is no way that we would know about this industry at Memphis, or indeed how the Roman occupation of the city changed the traditional methods of manufacture.

Stone Sculpture

A willingness on behalf of the rulers during the Ptolemaic and early Roman periods to reach both Greek and Egyptian audiences by promoting

Figure 5.4 • Almost complete faience bowl with a double outer ring and a small base ring; the glaze is light blue and unevenly applied with a thick accumulation at the bottom of the inside base (UC 49991) (© Petrie Museum of Egyptian Archaeology, University College London).

essentially 'bilingual' images of themselves is well attested (Ashton 2001: 26–35). Citing Thompson's study of the literary and textual material at the site, Smith (1996:210) suggested that the Egyptian royal statues with Greek portrait features may have originated at Memphis. He concluded: 'These images may be seen as the visual counterpart of the policy of accommodation with the Macedonian rulers pursued by the Memphite clergy through the medium of the royal cult'. Evidence from Petrie's excavations supports the idea that Greek and Egyptian artists were working side by side and that the latter in particular were influenced by the art of the former. A full synthesis of styles, however, does not occur until the Roman occupation of Egypt (Ashton 2003b:29–38).

Plaster and limestone models of rulers illustrate that there were royal sculpture workshops at Memphis, probably based in the area today known as Mit Rahina. The models are not, however, Greek in style (as one might expect if Greek portraits were copied by Egyptian sculptors); with the exception of one model, all the Petrie material can be dated to the 3rd century B.C.E. and is purely Egyptian in style. I define Egyptian as stylised/formal representations typical of Egyptian sculpture during this period and continuing the tradition of the 30th Dynasty (Ashton 2003b:60–63).

These 'portraits' can be divided into two separate categories: those that terminate in a bust form (which are typically stone) and those that show only the face of the ruler (for example, UC 33422 and UC 33423). More solid examples of the bust type of royal portrait have also been found at the site, but none are housed in the Petrie Museum. Lembke suggested that such models were used by the sculptors who carved the avenue of sphinxes, which originally led to the temple of Nectanebo at the Sarapieion (Lembke 1998:267–73). It is frustrating that Petrie does not mention the find spot(s) for the plaster and limestone material, but fortunate that he marked at least some examples with 'Memphis'. The material is enough to show that there was a royal sculptor on the site, but from the evidence it would seem that this was an early Egyptian workshop. Some of the plaster figures show parts of Egyptian-style representations, which may have been used as sculptors' studies for bronze or stone statutes. They were mould-made, and, on some, the details were added later; UC 33290 and UC 33291 show the striated kilt of a ruler or god. There are also examples of relief trial pieces of both royal and divine subjects, which suggests that temple-artists were also active at the site. Examples such as UC 33419 (Figure 5.5) show only part of the headdress and ear of a royal

Figure 5.5 • Limestone trial-piece with sculpted ear, side of face, and part of headdress of a king (UC 33419) (© Petrie Museum of Egyptian Archaeology, University College London).

subject, suggesting that some at least were trial or student pieces rather than models; this, in turn, has important implications for our knowledge of workshops at Memphis.

Smith (1996) was quite correct in concluding that the right cultural ingredients were found at Memphis, but it is really only when the non-royal terracotta figures are considered that we find more concrete examples of Greeks and Egyptians working side by side. Egyptian terracottas are traditionally hand-modelled; during the Ptolemaic period, however, a new form occurs that is pressed into a mould to form the figures but is left unmodelled and flat at the back (UC 30186:Figure 5.6; Ashton 2003b:73). Greek artists at Memphis may have influenced this innovation; the subjects are often phallic, and the example illustrated here was found at the Sacred Animal Necropolis at Saqqara; there are many examples of this type in the Petrie Museum and other collections. Török (1985:52–55, nos. 48–50) dates this form to the early Ptolemaic period, which would seem sensible when the iconography and manufacturing techniques are considered; other scholars (Fischer 1994:113–16, nos. 1–11) date this form to the 5th century B.C.E., although it is not clear why.

A reverse cross-cultural exchange of methods also occurs during the early Ptolemaic period and may have been responsible for a genre of Greek

Figure 5.6 • Ceramic plaque with female figure in relief (UC 30186) (© Petrie Museum of Egyptian Archaeology, University College London).

terracotta representations that spread throughout the Hellenistic world: the so-called grotesque figurines (Figure 5.7). The bodies of many of these figures are distorted, often contorted as if dancing, which led Török to suggest that they may represent dwarfs and were thus connected to the pharaonic tradition of performing dwarfs (Török 1985:138–39). The link in terms of manufacturing techniques to the Egyptian tradition may support this hypothesis. There are many bald-headed hand-modelled Egyptian-style heads which, like the Greek terracotta heads with exaggerated features, are solid (as was the Egyptian tradition). Furthermore, the Egyptian examples have swollen skulls, as was typical of the sculpture in the Ptolemaic period (Bianchi 1988:59; Bothmer 1960:100–01 and 110–11). Although Petrie originally suggested that these bald-headed hand-modelled terracotta heads represented foreigners, it has recently been suggested that the heads were from terracotta representations of priests (Ashton 2003a). In a book on realism in Alexandria, Himmelmann (1983) suggested that the so-called Hellenistic grotesques may have originated from Alexandria. Memphis, however, would seem to be a more logical place for the development of this form of figure because the solid heads and bodies are essentially an Egyptian phenomenon; in Greece, prior to the Hellenistic period, terracottas tend to be hollow and mould-made. The

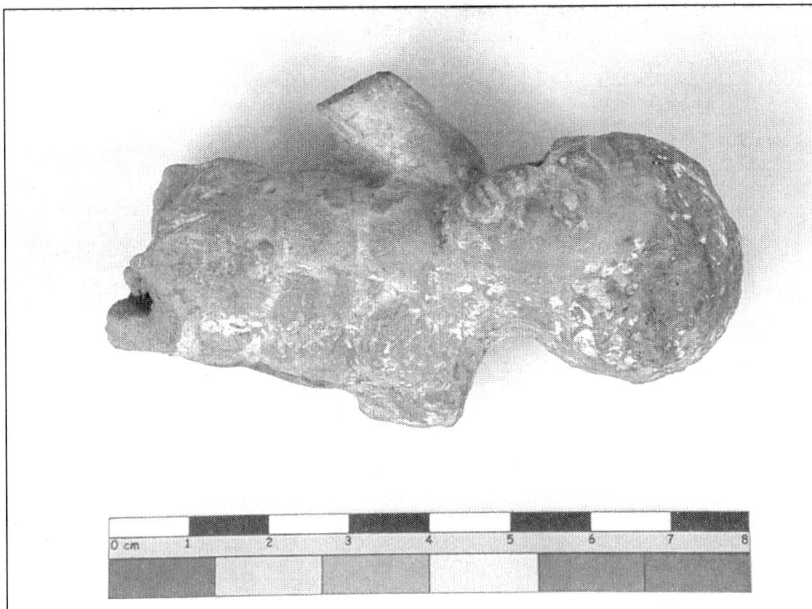

Figure 5.7 • Example of 'grotesque' terracotta figurine (UC 48250) (© Petrie Museum of Egyptian Archaeology, University College London).

terracotta finds support this hypothesis in that they show Greek and Egyptian workmen in close proximity of each other, perhaps suggesting a dissemination of ideas and methods of manufacture.

Some heads fall between Greek and Egyptian forms, again suggesting cross-cultural exchange. UC 48515 (Figure 5.8) may have been modelled by a Greek or an Egyptian, since it is close to the sculpted portraits found on Egyptian sculpture in the late 2nd century B.C.E. (compare Ashton 2001:56, 1.5). A further possible connection with sculpture is provided by the idea that Greek-style terracottas acted as models for artists (Conelly, in Uhlenbrock 1990:99), which would accord with the observed connection between stone sculpture and terracotta. The Petrie Museum material offers a unique insight into the working practices of artists at Memphis, and also identifies the city as an important artistic centre.

Bronze and Terracotta Figurine Production

We know that both terracotta and bronze figurines were produced at Memphis because of plaster moulds that were found there (Ashton 2003b: 67–68. Unfortunately, Petrie did not specify where this material was found, and, unusually for someone who was so interested in production,

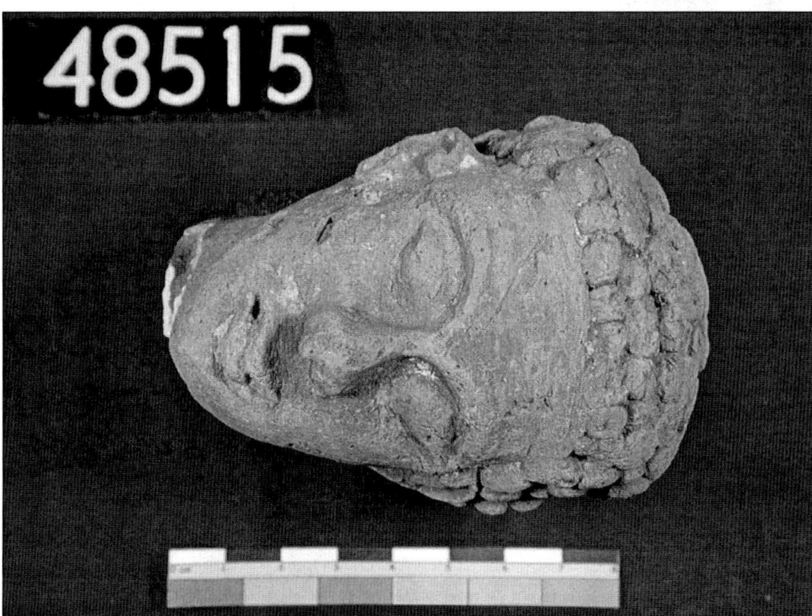

Figure 5.8 • Solid terracotta head of a male (UC 48515) (© Petrie Museum of Egyptian Archaeology, University College London).

he did not even publish images of the material. Like the royal plaster and stone models, some examples have 'Memphis' written on them, and so we can assume that they came from the site. In 1903, Edgar published the plaster moulds that were sent from Memphis to the Egyptian Museum in Cairo, although the majority of these were Greek in style; those at the Petrie Museum include both Greek and Egyptian examples. The plaster matrix of these two styles is different: those from Egyptian-style objects are made from a more compact form of plaster. Lucas, who wrote a manual on Egyptian technology and who worked with Edgar, may not have tested this type for Edgar's publication (Edgar 1903:iii), because the two types of plaster to which he refers were essentially the same.

The moulds appear to have been used for terracotta figurines, clay lamps, and bronze statuette manufacture. Egyptian terracotta figurines tend to be hand-modelled rather than mould-made, and so direct evidence of the manufacture of this group is not found at the site, and it can be presumed that the many composite moulds that are Egyptian in style were intended for bronzes. Similarly, Hellenistic Greek lamps are wheel-made rather than mould-made, and there is no evidence for their manufacture at the site. Edgar concluded that the majority of the Greek moulds in his publication were for bronze casting, and the pouring holes found on these examples are also clear on many of those in the Petrie Museum. Moulds were made in two halves, which were joined, and then the hot metal was poured into the centre to solidify as shown by UC 50219. The Petrie moulds fall into two main groups: Greek fittings and Egyptian royal and divine statuettes; the latter category, as illustrated by UC 50195 and UC 50220 (Figure 5.9), supports the idea that there were royal workshops at the site. Finally, moulds for Roman faience bowls were also found and provide further knowledge of the industry, along with the many kiln wasters. That none were found at Kom Helul, however, raises serious questions about the site being the original place of manufacture and support Petrie's observations regarding the material being dumped there. The new excavations at the site seem to indicate that the kilns were clean, which poses questions about their actual function.

In 1978, Roman pottery kilns were discovered in the courtyard of the so-called Hathor temple (El-Sayed Mahmud 1978:4; Jeffreys 1985:18); archaeology had revealed another industry that was unmentioned in the written record. As Thompson notes (1988:69), one would expect in a city the size of Memphis a certain degree of self-sufficiency in trades such as pottery production. Examples of partially manufactured or waste products for some object categories such as woodcarving and textile manufacture are not represented in the archaeological record (Thompson 1988:46–59, 70),

Figure 5.9 • Fragment of a plaster mould with an impression of the back of a nemes headdress with tail and shoulders (UC 50220) (© Petrie Museum of Egyptian Archaeology, University College London).

but they are mentioned in the papyri. This demonstrates that both archaeological and textual evidence should be used together in order to obtain the widest picture of any site. Until recently, at Memphis the material culture and technological evidence from Petrie's excavations had been neglected, but exactly what does a review of this evidence allow us to conclude?

Assessment of the Archaeological Evidence

The technological material from Memphis is not fully understood, and in order to obtain further information, particularly with regard to the differences between Greek and Egyptian, and Greek and Roman manufacturing techniques, further scientific testing is required. Ptolemaic and Roman faience is visually quite different and datable by form, but we do not know exactly how the components of faience were adapted in the early 1st century C.E. to produce such a different product. Similarly, further testing of the composition of Greek, Roman, and Egyptian plasters is necessary. For the most part, a detailed knowledge of Greek, Roman, and Egyptian material culture is necessary to date all the material from Petrie's excavations, and without this, given the lack of stratigraphic recording at the site, the only way to date the material is to use a knowledge of sculptural and ceramic forms. Much of the material was dumped or constituted surface finds, and this is also true of many of the comparative terracotta figures that were recorded at Saqqara (Martin 1981:nos. 371, 800). Because of Petrie's excavation methods, surface finds take on an equal importance to the objects from stratified deposits, and they are also important for our knowledge of the site. With this in mind, we might consider such material more carefully rather than dismissing it simply because it is without a 'real' context.

With a basic knowledge of the types of finds from the site and their respective dates, Petrie's material offers an illuminating window on artistic practices in Ptolemaic and Roman Memphis. The links between the Egyptian and Greek terracottas, the similarity between terracotta and stone workshops, and also the subjects that the artists decided to produce all allow the modern scholar an insight into a thriving multicultural society, which accords with Strabo's description of Memphis in the late 1st century B.C.E. In terms of a production centre, however, it seems unlikely that Memphis played second to Alexandria, and perhaps when we are searching for a centre of Greek artistic innovation we should be talking about a Memphite rather than an Alexandrian style.

Note

1. Based on observations by the author on the material she was permitted to see by Nicholson following the 2000 season. See also Nicholson (2001:15–17).

References

Ashton, S-A. 2001. *Ptolemaic Royal Sculpture from Egypt: The Interaction between Greek and Egyptian Traditions.* BAR International Series, 923. Oxford: Archaeopress.

———. 2003a. 'Foreigners at Memphis? Petrie's racial types', in W. J. Tait (ed.), *Never Had the Like Occurred: Egypt's View of Its Past.* London: University College London, 187–96.

———. 2003b. *Petrie's Ptolemaic and Roman Memphis.* London: Institute of Archaeology.

Bianchi, R. S. (ed.). 1988. *Cleopatra's Egypt: Age of the Ptolemies.* Exhibition Catalogue. Brooklyn, NY: The Brooklyn Museum.

Bothmer, B. V. 1960. *Egyptian Sculpture of the Late Period, 700 B.C. to A.D. 100.* Brooklyn, NY: The Brooklyn Museum.

Edgar, C. C. 1903. *Greek Moulds.* Cairo: Catalogue General du Caire.

El-Sayed Mahmud, A. 1978. *A New Temple for Hathor at Memphis.* Warminster: Aris & Phillips.

Fischer, J. 1994. 'Griechisch-Römische Terrakotten aus Ägypten', *Tübinger Studien 14 zur Archäologie und Kunstgeschichte.* Berlin: Ernst Wasmuth Verlag.

Himmelmann, N. 1983. *Alexandria und der Realismus in der Griechischen Kunst.* Tübingen: Verlag Ernst Wasmuth.

Jeffreys, D. G. 1985. *The Survey of Memphis 1.* London: Egypt Exploration Society.

Lembke, K. 1998. 'Die Sphingenallee von Saqqara und ihre Werkstatt', *Mitteilungen des Deutschen Archäologischen Instituts Abteilung Kairo* 54:267–73.

Martin, G. T. 1981. *The Sacred Animal Necropolis at North Saqqara.* London: Egypt Exploration Society.

Nicholson, P. T. 1998. 'Materials and technology', in F. D. Friedman, *Gifts of the Nile: Ancient Egyptian Faience.* London: Thames and Hudson, 50–64.

———. 2001. 'Faience production at Kom Helul, Memphis', *Egyptian Archaeology* 18:15–17.

Petrie, W. M. F. 1909. *Memphis I.* London: British School of Archaeology in Egypt.

———. 1911. 'The pottery kilns at Memphis', in E. B. Knobel, W. W. Midgley, J. G. Milne, M. A. Murray, and W. M. F. Petrie (eds.), *Historical Studies.* London, British School of Archaeology in Egypt and Bernard Quaritch, 34–37.

Smith, R. R. R. 1996. 'Ptolemaic portraits: Alexandrian types, Egyptian versions', in *Alexandria and Alexandrianism.* Papers delivered at a symposium organized by The J. Paul Getty Museum and The Getty Center for the History of Art and the Humanities and held at the museum, April 22–25, 1993. Malibu, CA: J. Paul Getty Museum, 203–14.

Stewart, A. 1996. 'The Alexandrian Style: A Mirage?' *Alexandria and Alexandrianism.* Papers delivered at a symposium organized by the J. Paul Getty Museum and the Getty Center for the History of Art and the Humanities and held at the museum, April 22–25, 1993. Malibu, CA: J. Paul Getty Museum, 231–46.

Thompson, D. J. 1988. *Memphis under the Ptolemies.* Princeton, NJ: Princeton University Press.

Török, L. 1985. 'Hellenistic and Roman terracottas from Egypt', in *Bibliotheca Archaologica 15: Monumenta Antiquitatas Extra Fines Hungariae Reperta* (Vol. 4). Rome: L'Erma.

Uhlenbrock, J. P. 1990. *The Coroplast's Art.* New York: Aristide D. Caratzas.

CHAPTER 6

Theophilus and the Use of Beech Ash as a Glassmaking Alkali

CAROLINE M. JACKSON AND JAMES W. SMEDLEY

ABSTRACT *The use of wood ash in glassmaking between the 10th and 12th centuries in Northern Europe heralded a significant change in technology, not only in the raw materials used to produce glass but also in the manufacturing locations, furnace design, and products manufactured within the region. Our earliest surviving description of this later glass production is that given by Theophilus in* De diversis artibus, *dated to the early 12th century. This account of medieval glass production has become widely accepted within the archaeological literature, which has established beechwood ash as the primary source of alkali for glassmaking in medieval Northern Europe. This chapter reviews some of the evidence that has led to the acceptance of the use of beechwood as the first choice of alkali for Northern European glassmaking by examining the contemporary documentary texts alongside other archaeological, environmental, analytical, and experimental evidence. It is shown that this integrated approach leads to a revised history of medieval glass technology.*

Forest Glass and the Use of Beech Ash in Glassmaking

Manmade glass is produced by fusing, at high temperatures, silica, a flux, and a stabiliser (Turner 1956). In archaeological glasses, the silica is derived from quartz pebbles or silica sand; the flux, containing an alkali at high concentrations, from a soda-rich evaporate such as natron or from plant ashes. Calcium acts as a stabiliser in glasses, preventing the glassy network from disintegration. It is unclear how this was introduced to the glass; whether intentionally through the use of a calcium-rich mineral such as dolomite, or unintentionally through impure silica sands or plant ashes. Other substances, usually in mineral form or derived from the glass-forming raw materials, were introduced to the melt to colour, decolour, or opacify the glass (Freestone, this volume).

The use of potassium-rich woodland plants to produce ash for glass manufacture between the 10th and 12th centuries in Northern Europe heralded a change, not only in the raw materials used to produce glass but also in the manufacturing locations, furnace design, and products found and manufactured within the region. Before this change, the archaeological evidence suggests that glass was made using alkali raw materials which were soda-rich, derived either from evaporite deposits or salt-loving plants such as *Salicornia* or *Salsola*. When heated with silica, these plant ashes produce glass compositions that are high in sodium and relatively low in potassium. The location of manufacture for many of these early glasses is unclear, but the prevailing hypothesis is that raw glass may have been produced in large primary installations and shipped for reworking elsewhere, or later was produced from recycled glasses (Brill 1988; Callmer and Henderson 1991; Foy, Vichy, and Picon 2000; Freestone, Greenwood, and Gorin-Rosen 2002; Nicholson, Jackson, and Trott 1997; Price 1978; Rehren and Pusch 1997). The shift to potash glass in the Middle Ages, using wood ash alkalis, placed glass production in a number of small primary locations throughout Northern Europe (for example, see Foy 1991; Kenyon 1967; Krueger 1996).

Our earliest surviving description of this later glass production is that given by Theophilus in *De diversis artibus*, dated to the early 12th century (Hawthorne and Smith 1979), a manuscript that is thought to be broadly contemporary with this change in glassmaking practices. Theophilus, who may have been Roger of Helmarshausen, a German lay brother with practising knowledge of metalworking, wrote his treatise in three parts; painting, glass, and metalworking. In Book Two, *The Art of the Worker in Glass*, Theophilus describes the raw materials used to manufacture glass. Of interest here, he writes that beech was the alkali used to manufacture potash-rich forest glass.

> If you have the intention of making glass, first cut many beechwood logs and dry them out. Then burn them all together in a clean place and carefully collect the ashes, taking care that you do not mix any earth or stones with them [. . .] (Hawthorne and Smith 1979:49)

Theophilus also suggests that beech logs were used for fuel to heat the furnace: 'take beechwood logs completely dried out in smoke, and light large fires in both sides of the bigger furnace' (Hawthorne and

Smith 1979:52). In this case, the two uses are discussed separately, which implies that the fuel ash was not used as a source of alkali. However, a *specific* wood has been chosen for fuel that, whether by accident or design, could be used as an excellent source of alkali for glassmaking. In other medieval and postmedieval contexts, the use of fuel ash as a supplementary source of alkali has been suggested (Crossley 1967, 1990; Hurst Vose 1980; Kenyon 1967).

It is these statements that have become widely accepted within the literature and that have placed beechwood ash as the central and primary source of alkali for glassmaking in medieval Northern Europe (for example, Freestone 1992; Hartmann 1994; Newton 1980; Newton and Davison 1989:57; Smrček 1996). This chapter reviews some of the evidence that has led to the acceptance of the use of beechwood as the first choice of alkali for Northern European glassmaking by examining the contemporary documentary texts alongside other archaeological, vegetational, analytical, and experimental evidence.

The Texts

Although Theophilus is the most famous source, a number of other manuals discuss the use of beech ash in glassmaking, including those attributed to Agricola (Hoover and Hoover 1950) and Eraclius (Merrifield 1967). The work of Georgius Agricola, translated from the 1556 edition, also describes glassmaking in Germany. He mentions beech, not as a primary choice of alkali, as Theophilus suggests, but as a substitute if other materials were unavailable.

> When the said juices [solidified juices from soda, rock-salt or herb salts] are not to be had, two parts of the ashes of oak or holmoak, or of hard oak or turkey oak, or if these be not available, of beech or pine, are mixed with one part of coarse or fine sand [. . .] The ashes should be made from *old* trees, of which the trunk at a height of six feet is hollowed out and fire is put in, and thus the whole tree is consumed and converted into ashes. (Hoover and Hoover 1950, 586)

A variation in recipe and use of beech in its *early* growth is discussed by Merrifield (1967) in her translation of the works of Eraclius. This manuscript may have been contemporary with that of Theophilus and is thought to have been written in the 12th or 13th centuries but probably relates to Italian glassmaking (Merrifield 1967:167). Eraclius, in his recipe for white glass and glass of various colours, states: 'Glass is made with ashes both of fern and of "faina" – that is, of the small trees which grow in the woods'. Merrifield (1967:212) suggests these latter ashes may be produced from the seedlings of wild beech trees. Further 'then take two parts of fern, and one-third part of faina and mix them together'. Moreover: 'Purple and "*membranaceum*" [glass] are made differently, with the ashes of the beech-tree, which are baked like white ashes, and put into a melting pot, and melted and boiled until they burn to a purple colour'.

Three important points emerge from the comparison of these three documents. The first is that both Theophilus and Agricola suggest the use of mature beech trees, whereas Eraclius mentions saplings. This has important consequences when one considers the chemical composition of the ashes produced by burning different parts of the same tree and in turn for forestry management (Geilmann 1955; Sanderson and Hunter 1981; Smrček 1996; Tichane 1998:25). The second is that, in some cases, ashes (including, but not primarily, beech) were mixed together in glass manufacture, and the third that certain species were chosen for specific glasses, in particular, for coloured glasses. This point is discussed in the manuscripts of both Theophilus and Eraclius (Hawthorne and Smith 1979:57; Merrifield 1967:212).

This last point has a specific importance for the use of beech ashes for glass production in the Middle Ages. A significant proportion of glass at this time was manufactured for stained glasses for use in ecclesiastical buildings. The colour of the glass was obviously of great importance, and although some glasses were painted, others were coloured either by the addition of metals such as copper to the glass melt, or were naturally coloured with the use of specific raw materials. Beech ashes fall into this latter category, since they often contain significant proportions of manganese and iron that, under optimum melting redox conditions and in the absence of other strong colouring agents, turn the glass purple (oxidising) or yellow (reducing) (Geilmann and Brückbauer 1954; Sellner, Oel, and Camera 1979). Both Theophilus and Eraclius were aware of this quality in beech ashes and specifically mention this in their treatises.

Neri (1612) does not mention beech in his writings but lists a wide variety of other plants, including beans, bramble, millet, rushes, and fern (Cable 2001:71–73) in addition to imported polverine and rochetta (Cable 2001:59). However, Merret, in his substantial addition to Neri's work, published in 1662, mentions beech in passing: first, in the context of the writings of Agricola, which Merret quotes, and second in his observations concerning the production of 'Crystall [*sic*] for making of jewels' (Cable 2001:378). Again, beech is chosen because it can produce a wide variety of colours in the finished glass:

> Dissolve, saith he, in water two ounces of purified salt of Tartar, which moisten with Beechen-ashes make thereof balls as big as apples; Dry and burn them in a potters furnace in a covered pot, for so the balls with somewhat melt, and stick to one another, let them then be finely pounded and a Lee made thereof, which congele to a Salt [. . .] Take then of purified salt of Tartar two parts, of the foregoing salt preparted, one part, melt them together. This will receive all the colours of the whole world, and appear like Oriental gems. (Cable 2001:379)

The documentary evidence discussed above is a convincing testimony that beech ash was widely known as a glassmaking alkali by a number of authors. We do not know how widely it was used. Therefore, before we accept this as the primary alkali source, we should investigate other evidence, in particular, any experimental evidence that supports the observations documented in these manuscripts or compositional trends in glasses of this period that could indicate the use of beech in glass manufacture. It is only through complementary documentary, experimental, and analytical evidence that we can gain a greater insight and understanding of the use of beech ash as a glassmaking alkali.

Archaeological: Experimental and Analytical Evidence

Two promising sources of evidence for the use of specific raw materials in glass manufacture come from experimental replication of glasses and the analyses of archaeological glass compositions. Experimental glasses, using beech ash, have been made by a number of authors, in attempts to replicate the recipes described by Theophilus (Royce-Roll 1994; Smedley, Jackson, and Booth 1998; Wedepohl 1997) and to ascertain the colouring properties of 'forest glass' under different redox environments (Geilmann 1955; Sellner, Oel, and Camera 1979). In all cases, beech ash (or replicates based on the components found in beech) and sand as raw materials produced glass at temperatures that could probably have been attained in medieval and postmedieval furnaces (Cable and Smedley 1986; Smedley, Jackson, and Booth 1998) and a wide variety of colours were achieved depending on the composition of the beech ash, the ratio of ash to sand, the time and temperature of melting, and the redox conditions (Geilmann 1955; Royce-Roll 1994; Sellner, Oel, and Camera 1979). Royce-Roll (1994) specifically noted the simplicity of the recipe and production process. However, although this experimental work showed that coloured glass can be produced using beech ash and sand, it does not prove that beech was used *widely* for glass manufacture in medieval and postmedieval Europe.

Analytical work on glasses from medieval and postmedieval Europe has often been tentative in attributing glasses to specific alkali sources such as beech. Many papers define their raw materials only to a generic *wood ash* (for example, Gerth, Wedepohl, and Heide 1998). The following section discusses a number of papers that do attempt to identify beech ash as a source of alkali through the chemical analyses of archaeological glasses. These are separated into three case studies relating to different areas of Northern Europe, including Germany, which is the home location of the beech ash glasses described by Theophilus and Agricola. In the 12th century, Theophilus suggests that glass in Germany was produced solely using beechwood ash. By the 16th century, Agricola describes the mixing of beech with other alkalis. However, we have no textual evidence that glasses produced in other parts of Northern Europe were also manufactured using the same alkali raw materials.

Case Study 1: Germany

The earliest wood-ash glasses in Germany, putatively produced from beech ash, are discussed by Hartmann and colleagues (1997). An analysis of green and turquoise ring beads from Kassel, in Germany, dated from the 12th–8th century B.C.E., displayed an unusual composition, with higher potash and lower soda values than would be expected in a glass produced from a salt-loving herb, such as *Salicornia*, which is more usual in glasses of this period. These compositions, they suggest, are indicative of glasses produced using a processed woodland plant ash, and their experimental evidence suggests this is most probably beech (Hartmann et al. 1997:554). At present, this is by far the earliest indication that beech was used in glass manufacture and is a singular example.

In his analysis of early and late medieval glasses from a number of sites in Germany, Wedepohl (1997) suggests that high lime to potash ratios in the early glasses he analysed, in the order of 1 to 1.5, indicate the mixing of high proportions of spruce to beech in manufacture, or alternatively indicate the use of ash derived from parts of the beech tree other than trunk wood. This particular composition would be produced using a glass batch formulation of ash to sand in the order of 1:1 to 2:1 (Wedepohl 1997:249). His assertion is that beech was an essential part of the glassmaking batch and that this was the major glass type in Germany from 1000–1400 C.E. This conclusion stems from earlier experimental work by Wedepohl and Hartmann (unpublished, quoted in Wedepohl 1997), but it also refers to Theophilus and his statements concerning the use of beech ash. Wedepohl suggests later glasses (post-1400 C.E.) were produced with wood ashes and sand, with the addition of lime to conserve alkali stocks. This is illustrated by Hartmann (1994) in his analysis of 15th- and 16th-century glass from the Eichsfeld in Germany; he suggests that the glass from this region is manufactured to a specific recipe of 20% limestone, 20% beechwood ash, and 60% quartz sand. The presence of manganese in these glasses is attributed to the use of beechwood (Hartmann 1994:113). The analysis is supported by the experimental production of glass using beechwood ash from the trunks and boughs of trees (although it is suggested oak ash could also provide a viable alternative).

Case Study 2: France

Barrera and Velde (1989) and subsequently Marquis and associates (2000) identified wood ash glasses in a very large suite of 13th–16th-century glasses analysed from France. Using published compositions of beech, oak, and fern ashes, the ratios of lime to lime + potash in these glasses led them to the conclusion that the majority of glasses were produced using mixed ash batches. They suggest that beech ash alone would be too rich in calcium to produce the observed ratios and that some of the glasses were probably made using a combination of beech and fern ash, or possibly oak and fern ashes, or even fern ash and oak or beech ash *extract* (Barrera and Velde 1989:94). This recipe mirrors the two component alkali batches discussed by Eraclius (Merrifield 1967:212). They also suggest that some glasses may have been produced using oak ashes because of the very high calcium content; however, Geilmann (1955) indicates that beech ash in certain circumstances can contain up to 64% lime. This complication highlights the difficulty of determining species, since it is clear plant ashes are compositionally variable in nature.

Case Study 3: Britain

Tennent and colleagues (1984) analysed a number of 13th-century stained glass fragments from Elgin Cathedral, Scotland, a building that was founded in 1224 and rebuilt in 1270. The source of the glass is unknown, and although stained glass was made in England in this period, they suggest it may equally have originated from France. Using the ratios of potash to magnesia in their glasses, in comparison with data from Turner (1956: 289T), they suggest that the glass was produced using a mixture of beech and fern ash, a mixture they note is mentioned in the texts of Eraclius. This conclusion must be tentative, since it uses one set of ash analyses per species and does not take into account the variability in raw materials. What this and the other papers discussed above illustrate is the way in which the medieval texts are used to confirm preliminary hypotheses about medieval glassmaking raw materials.

* * *

The cases discussed above highlight the difficulties in determining the species of wood used to provide the alkali component of medieval and later glasses. The main factor clouding the interpretation of these glass compositions is the wide variation in wood composition within and between species, which ultimately affects the composition of the ashes used in glassmaking (for example, Bezborodov 1975; Geilmann 1955; Hartmann 1994; Sanderson and Hunter 1981; Smrček 1996; Turner 1956). In nearly all the cases reviewed, this has led to the authors suggesting that a mixture of ashes were used for glassmaking, including beech as one component. The evidence presented suggests that, in isolation, no single line of evidence can *prove* the use of beech, but together they provide an indication for the use of particular plant ashes at differing locations throughout Northern Europe in the medieval and postmedieval periods. The discussion of this evidence brings us back to the central question of this paper: to what level of confidence can we suggest that beech ash was the primary alkali used in the Middle Ages in Northern Europe?

Beech as the Primary Wood Ash Alkali in Northern European Glassmaking

The use of a particular species of tree for the production of ash for glass manufacture presupposes the presence of each wood type in sufficient quantity within the glassmaking locality. Although the importation of raw materials from the Levant, specifically for the Venetian glass industry, has been discussed in the literature (Jacoby 1993; McCray 1999), one generally assumes that Northern European medieval and postmedieval glasses were produced using local raw materials. This is because of an increase in the number of glasshouses found in woodland areas at this time and the existence of small glass producing centres in some areas that were clearly in use for only short periods (Crossley 1990; Kenyon 1967). Although it is known that ashes were sold to glassmaking houses (McCray 1999; Smith 1962; Welch 1997), in most cases it can be suggested that these materials were not transported great distances. Therefore, if beech was used, one must assume that it was readily available, in quantity, within each of the glassmaking regions within Northern Europe.

Published findings have shown that the yield of ash from hardwoods is very low, generally in the order of less than 2% (Bezborodov 1975; Smedley, Jackson, and Booth 1998; Wedepohl 1997). Thus, to produce enough ash for glassmaking would have required large amounts of beech, even at such a small ratio of 2 parts ash to 1 part sand (Smedley and Jackson 2002a). Using such data, Wedepohl (1997:254) suggests that German glass houses had to move every 5–30 years because of their need for wood, for fuel, and alkali, and this requirement would have deforested an area of approximately 15 km^2 per year. These data reinforce the need for a sustainable yield of beech for alkali and imply the need for forest management. They may also explain the small size of most glassmaking facilities in woodland areas in the Middle Ages and the high proportion of furnace sites within a locality.

The evidence for abundant beech woodland in medieval Europe is mixed. The writings of Theophilus, along with pollen records of the time, indicate that beech was plentiful throughout Germany. Tyzak (1995:5) also places at least one 15th-century French glass house in the beech-wood of the Bois le Compte, although it is clear from contemporary documentary records that beech was not the sole alkali used in glass-making in this region at this time (Tyzak 1995:19). However, it is not clear that other glassmaking areas in Europe were so endowed. Using pollen records and historical sources, Newton (1980:178) suggests that beech trees were rare south of the Alps and that in the Surrey Weald, England, an area known to be producing glass in the medieval period, the woodland presently consists of up to 95% oak. Godwin (1975:273) proposes that this view may arise from the writings of Julius Caesar's, which said beech (*Fagus*) was not present in these woodlands in the *Roman* period. Godwin then goes on to indicate that the picture is more complex. First, although Caesar mentions the absence of *Fagus*, Godwin (1975:273) argues that it is likely that he was referring to sweet chestnut. Second, beech (*Fagus sylvatica*) is underrepresented in the pollen record because of the low productivity of beech pollen and because the pollen is large and heavy and does not circulate easily in air currents. This makes any estimation of its abundance difficult. The presence of beech in the Weald is supported by Mabey, who states that beech was used as a fuel for the Roman iron industry, and during the medieval period 'in the Weald, beech was the fuel of choice for the glass industry' (1996:79), indicating not only the presence of beech but also its abundance. The use of local beech billets for fuel in the later Wealden glass industry is supported by a quote from the glassmaker Henry Sturdwick of Kirdford in 1557, who leaves upon death 'the profit of my Glasse Howse, withal the beches that I have bought, and half the beches in and upon Idehurst and Crofts' to his sons (Kenyon 1967:45).

This apparent confusion concerning the presence of beech throughout medieval Britain places a question mark around its use as an alkali in glass manufacture throughout Northern Europe. If, as Newton suggests, beech was not the prominent tree in medieval woodlands, glasshouses south of the Alps and in southern England could either have brought in beech ashes or used alternative alkali sources. The latter seems more likely. However, even if beech was native by this period as Godwin (1975), tentatively, and Mabey (1996), definitely, suggest (and documentary evidence sustains), then the picture would indicate that beech could have been used as a glassmaking alkali south of the Alps. There is further supporting documentary evidence and analytical supposition that in some areas beech ashes were mixed (Barrera and Velde 1989; Merrifield 1967; Smedley and Jackson 2002b; Tennent et al. 1984; Tyzak 1995). The mixing of different hardwood ashes with fern, which has a high alkali content, would reduce the quantity of beech ash required and has been shown to benefit the melting performance of ash plus sand mixtures (Smedley, Jackson, and Welch 2003). Alternatively, there is equal weight to the use of alternative alkali sources such as oak (Barrera and Velde 1989; Hartmann 1994) or fern in later periods (Henkes 1994; Smedley and Jackson 2002b).

Discussion and Conclusion

The acceptance of beech as the main, and in some cases only, wood ash alkali in use in the medieval and later periods is questioned. Although documentary evidence is persuasive, beech being mentioned by a number of authors ranging from the 12th to 16th centuries, it does not preclude the use of other ashes, some as diverse as thistle and hawthorne. Some authors, such as Neri, omit any reference to the use of beech ash in glass manufacture.

Similarly analytical evidence has often been most tentative in its attribution to specific alkali sources because of a lack of understanding of the raw materials used. To this end, there is now a growing body of literature that attempts to characterise plant ashes from different species and regions (for instance, Brill 1999; Gerth, Wedepohl, and Heide 1998; Hartmann 1994; Jackson, Smedley, and Booth 2005; Sanderson and Hunter 1981). However, as the data set grows, the attribution of specific ash compositions to plant species becomes less clear. For example, whereas in some cases oak and beech can be compositionally distinguished Jackson and Smedley (2004), Sanderson and Hunter (1981), and Hartmann (1994) have illustrated that samples of beech and oak taken from neighbouring locations could not be successfully distinguished (Table 6.1). Similarly, the presence of relatively high levels of manganese in a wood-ash glass is often thought to indicate the use of beech ash. However, this element can also be present in other plant ashes such as oak, spruce, and bracken, or it can be added to glasses in mineral form to decolour or colour the glass, which makes its use as an indicator for beech ash use in glassmaking less secure. The characterisation of other species from differing locations is being undertaken by the authors and others (for example, Misra,

Table 6.1 Selection of compositions of beech and oak ash from different geographical regions.

	Beech Ash				Oak Ash
	Jackson and Smedley (2004) Sheffield, UK	Sanderson and Hunter (1981) Weald, UK	Sanderson and Hunter (1981) Weald, UK	Hartmann (1994) Lower Saxony	Hartmann (1994) Lower Saxony
SiO_2	18.03			0.9	3.9
Na_2O	0.59	2.35	3.02	0.08	0.49
K_2O	19.98	7.83	16.26	19.99	29.6
CaO	31.07	26.31	13.99	45.5	43.1
MgO	6.95	1.49	0.65	11.79	5.4
Al_2O_3	0.92			0.1	0.3
Fe_2O_3	0.94	0.43	0.22	0.47	0.77
P_2O_5	15.26			9.78	11.4
MnO	6.20	2.93	4.00	2.84	3.53

Ragland, and Baker 1993), and this extended database may help define the compositional parameters and tolerances of different ashes. Present results suggest that it is tentatively possible to distinguish between glasses produced from hardwoods and those produced from forest floor plants such as bracken; however, it is not yet possible to distinguish between, or even successfully identify, glasses made from beech and oak.

However, the characterisation of different ashes will only partly help identify the raw materials used in glass manufacture in the past, since the relationship between ash data and glass compositional data has to be treated with some degree of caution. There are a number of other natural variables that influence the composition of the glass. The variability of the same species of ash between and within geographical areas depending on substrate, the differences in composition of the ash from differing parts of the same tree, the compositional changes that take place because of seasonal changes in some species, and the effects of climate on vegetation have already been noted as problematic variables (Bezborodov 1975; Hartmann 1994; Turner 1956). However, other factors introduced during glass melting add to the difficulties associated with working from the finished glass to the raw materials used. These include the insolubility of some compounds in the glass melt, the volatilisation of others during heating, and the recipes used by the glassmakers (Jackson and Smedley 2004; Jackson, Smedley, and Booth 2005).

Thus, the use of beech as an alkali in glassmaking should be examined in context. The evidence presented above suggests beech was *one* type of glassmaking alkali used in the medieval period, but that many others were also used. The choice of alkali may have been affected by a myriad of factors that could have included geographical location and availability. Further work should combine historical research with vegetational history and glassmaking practices, alongside compositional analyses of historical glasses. However, the key to unlocking glass compositional data may lie with future work on the variation within raw materials and subsequent experimental work on the impact of different glassmaking protocols on glass compositions. It is this latter research that will expand our understanding of medieval and postmedieval glassmaking practices and allow us to put into place the use of beech as a glassmaking alkali.

Note

This paper was written in 2002. Any subsequent publications relating to medieval glassmaking or raw materials that may influence these conclusions could not be acknowledged or included.

References

Barrera, J., and B. Velde. 1989. 'A study of French medieval glass composition'. *Archéologie Médiévale* XIX:81–130.

Bezborodov, M. A. 1975. *Chemie und Technologie der antiken und mittelalterlichen Gläser*. Mainz: von Zabern.

Brill, R. H. 1988. 'Scientific investigations of the Jalame glass and related finds', in G. Weinberg (ed.), *Excavations at Jalame: Site of a Glass Factory in Late Roman Palestine*. Columbia: University of Missouri Press, 257–93.

———. 1999. *Chemical Analyses of Early Glasses* (Vols. 1 and 20). Corning, NY: Corning Museum of Glass.

Cable, M. 2001. *The World's Most Famous Book on Glassmaking: The Art of Glass, by Antonio Neri and Christopher Merret*. Sheffield: Society of Glass Technology.

Cable, M., and J. W. Smedley. 1986. 'Liquidus temperatures and melting characteristics of some early container glasses', *Glass Technology* 28(2):94–98.

Callmer, J., and J. Henderson. 1991. 'Glassworking at Ahus, S. Sweden (eighth-century AD)', *Laborativ Archeologi* (Archaeological Research Laboratory Stockholm University) 5:143–54.

Crossley, D. W. 1967. 'Glassmaking in Bagot's Park', *Post-Medieval Archaeology* 1:44–83.

———. 1990. *Post-Medieval Archaeology of Britain*. Leicester: Leicester University Press.

Foy, D. 1991. 'Ateliers de verriers de l'Antiquité et du haut Moyen Age en France: Méthodologie et résultats, un état de la question (enquête reprise et mise à jour, à partir du catalogue de l'exposition "A travers le verre, du Moyen Age à la Renaissance")', in D. Foy and G. Sennequier (eds.), *Ateliers de Verriers de l'Antiquité à la Période Pré-Industrielle*. Rouen: Actes des Journées d'etude Congrès d'AFAV, November 1989, 55–73.

Foy, D., M. Vichy, and M. Picon. 2000. Lingots de verre en Méditerranée occidentale (IIIe siècle av. J.-C. - VIIe siècle ap. J.-C.), *Annales du 14e Congrès de l'Association Internationale pour l'Histoire du Verre, Venezia-Milano 1998*, AIHV, Lochem. 51–57.

Freestone, I. C. 1992. 'Theophilus and the composition of medieval glass', in P. B. Vandiver, J. R. Druzik, G. S. Wheeler, and I. C. Freestone (eds.), *Materials Issues in Art and Archaeology III.* Warrendale, PA: Materials Research Society, 739–46.

Freestone, I. C., R. Greenwood, and Y. Gorin-Rosen. 2002. 'Byzantine and early Islamic glassmaking in the eastern Mediterranean: Production and distribution of primary glass', in G. Kordas (ed.), *Hyalos, Vitrum, Glass.* 1st International conference., Glasnet Publications, Athens.

Geilmann, W. 1955. 'Die chemische Zusammensetzung einiger alter Gläser, insbesondere deutscher Gläser des 10. bis 18. Jahrhunderts', *Glastechnische Berichte* 28(4):146–56.

Geilmann, W., and T. Brückbauer. 1954. 'Beiträge zur Kenntnis alter Gläser II: Der Mangangehalt alter Gläser', *Glastechnische Berichte* 27(12):456–59.

Gerth, K., K. H. Wedepohl, and K. Heide. 1998. 'Experimental melts to explore the technique of medieval woodash glass production and chlorine content of medieval glass types', *Chemie der Erde* 58:219–32.

Godwin, H. 1975. *The History of the British Flora.* Cambridge: Cambridge University Press.

Hartmann, G. 1994. 'Late-medieval glass manufacture in the Eichsfeld region (Thuringia, Germany)', *Chemie der Erde* 54:103–28.

Hartmann, G., I. Kappel, K. Grote, and B. Arndt. 1997. 'Chemistry and technology of prehistoric glass from Lower Saxony and Hesse', *Journal of Archaeological Science* 24:547–59.

Hawthorne, J. G., and C. S. Smith. 1979. *Theophilus, on Divers Arts.* New York: Dover.

Henkes, H. H. 1994. *Glas zonder Glans/Glass without Gloss: Utility Glass from Five Centuries Excavated in the Low Countries 1300–1800.* Rotterdam Papers, 9. Rotterdam: Coördinatie Commissie van Advies inzake Archeologish Onderzoek binnen het Ressort.

Hoover, H. C., and L. H. Hoover. 1950. *Georgius Agricola, De Re Metallica.* New York: Dover.

Hurst Vose, R. 1980. *Glass.* London: Collins Archaeology.

Jacoby, D. 1993. 'Raw materials for the glass industries of Venice and the Teraferma, about 1370–1460', *Journal of Glass Studies* 35:65–90.

Jackson, C. M. and J. W. Smedley. 2004. 'Medieval and post-medieval glass technology: Melting characteristics of some glasses melted from vegetable ash and sand mixtures', *Glass Technology* 45(1):36–42.

Jackson, C. M., J. W. Smedley, and C. A. Booth. 2005. 'Glass by Design? Raw materials, recipes and compositional data', *Archaeometry* 47(4):781–95.

Kenyon, G. H. 1967. *The Glass Industry of the Weald.* Leicester: Leicester University Press.

Krueger, I. 1996. 'Research in medieval glass: Where are we standing now?', *Annales du 13e Congres de l'association Internationale pour l'Histoire du Verre*, 277–88.

Mabey, R. 1996. *Flora Britannica.* London: Sinclair-Stevenson.

Marquis, P., M. T. Penna, C. Riquier-Bouclet, and B. Velde. 2000. 'Late medieval and Renaissance glassware from the rue des Lombards, Paris', *Journal of Glass Studies* 42:97–112.

McCray, P. 1999. *Glassmaking in Renaissance Venice.* Brookfield, VT: Ashgate.

Merrifield, M. P. 1967. *Medieval and Renaissance Treatises on the Arts of Painting.* New York: Dover.

Misra, M. K., K. W. Ragland, and A. J. Baker. 1993. 'Wood ash composition as a function of furnace temperature', *Biomass and Bioenergy* 4(2):103–16.

Newton, R. G. 1980. 'Recent views on ancient glasses', *Glass Technology* 21(4): 173–83.

Newton, R. G., and S. Davison. 1989. *Conservation of Glass.* Oxford: Butterworths.

Nicholson, P. T., C. M. Jackson, and K. M. Trott. 1997. 'The Ulu Burun glass ingots, cylindrical vessels and Egyptian glass', *Journal of Egyptian Archaeology* 83:143–53.

Price, J. 1978. 'Trade in glass', in J. du Plat Taylor and H. Cleere (eds.), *Roman Shipping and Trade: Britain and the Rhine Provinces.* CBA Research Report 24, 70–78.

Rehren, T., and E. B. Pusch. 1997. 'New Kingdom glass melting crucibles from Qantir-Piramesses'. *Journal of Egyptian Archaeology* 83:127–41.

Royce-Roll, D. 1994. 'The colors of romanesque stained glass', *Journal of Glass Studies* 36:71–80.

Sanderson, D., and J. R. Hunter. 1981. 'Compositional variability in vegetable ash', *Science & Archaeology* 23:27–30.

Sellner, C., H. Oel, and B. Camera. 1979. 'Untersuchung alter Gläser (Waldglas) auf Zusammenhang von Zusammensetzung, Farbe und Schmelzatmosphäre mit der Elektronenspektroskopie und der Elektronenspinresonanz (ESR)', *Glastechnische Berichte* 52:255–64.

Smedley, J. W., and C. M. Jackson. 2002a. 'Medieval and post-medieval glass technology: Batch measuring practices', *Glass Technology* 43:22–27.

———. 2002b. 'Medieval and post-medieval glass technology: A review of bracken in glassmaking', *Glass Technology* 43:1–4.

Smedley, J. W., C. M. Jackson, and C. A. Booth. 1998. 'Back to the roots: The raw materials, glass recipes and glassmaking practices of Theophilus', in P. McCray and W. D. Kingery (eds.), *The Prehistory and History of Glass and Glassmaking Technology. Ceramics and Civilisation: Volume VIII.* Columbus, OH: American Ceramic Soaity, 145–65.

Smedley, J. W., C. M. Jackson, and C. Welch. 2003. 'Unravelling glass compositions: Detecting the use of specific glassmaking raw materials at Little Birches, Staffordshire', *Annales du 15e Congrès de l'Association Internationale pour l'Histoire du Verre,* 203–07.

Smith, R. S. 1962. 'Glassmaking at Wollaton in the early seventeenth century', *Transactions of the Thoroton Society of Nottingham* 66:24–34.

Smrček, A. 1996. 'Sklářésuroviny a Kmeny v Minulosti. I. Základní Suroviny'. *Sklář a Keramik* 46(4):79–88.

Tennent, N. H., P. McKenna, K. K. N. Lo, G. McLean, and J. M. Ottaway. 1984. 'Major, minor and trace element analysis of medieval stained glass', in J. B. Lambert (ed.), *Archaeological Chemistry III,* Advances in Chemistry 205. Washington, DC: American Chemical Society, 133–50.

Tichane, R. 1998. *Ash glazes.* Iola, WI: Krause Publications.

Turner, W. E. S. 1956. 'Studies in ancient glasses and glassmaking processes: Part V, Raw materials and melting processes', *Journal of the Society of Glass Technology* 40:277–300.

Tyzak, D. 1995. *Glass, Tools and Tyzaks.* Buckinghamshire: Don Tyzak.

Wedepohl, K. H. 1997. 'Chemical composition of medieval glass from excavations in West Germany', *Glastechnische Berichte* 70(8):246–55.

Welch, C. M. 1997. 'Glassmaking in Wolseley, Staffordshire', *Post-Medieval Archaeology* 31:1–60.

CHAPTER 7

Medieval Precious Metal Refining: Archaeology and Contemporary Texts Compared

JUSTINE BAYLEY

ABSTRACT *The examination of metalworking debris from sites of medieval date in Britain has led to the identification of evidence for a variety of precious metal refining processes: cupellation (the separation of precious metals from base metals), assaying (testing the purity of gold and silver), salt parting, and acid parting (two methods of separating gold from silver). Initially, the finds posed problems of interpretation and their functions were unclear. This chapter shows how by marrying the descriptions contained in medieval texts with chemical and mineralogical analyses of the finds, the archaeological remains can be linked to particular processes, resolving many of the original uncertainties. Furthermore, some explanations are proposed to the discrepancies identified between historical and archaeological information. The relative value of both types of sources, and the potential of their combination, is addressed.*

Introduction

The precious metals of the medieval period were the same as today, gold and silver. They were used for making jewellery and decorative objects, and for coinage – and, again like today, they were often debased. Knowing their purity was important to those who used them, so assaying and refining processes were a significant part of medieval metalworkers' skills. The examination of metalworking debris from sites of medieval date in Britain has led to the identification of evidence for a variety of precious metal refining processes: cupellation (the separation of precious metals from base metals), assaying (testing the purity of gold and silver), salt parting, and acid parting (two methods of separating gold from silver) (for example, Bayley 1991a, 1991b, 1993; Bayley and Eckstein 1997, 2006). Initially, the finds posed problems of interpretation, and their functions were unclear; for example, ceramic cupels or scorifiers were described as 'heating trays' (for instance, Roesdahl 1977). However, by marrying the descriptions contained in medieval texts with chemical and mineralogical analyses of the finds, they can be linked to particular processes, resolving many of the original uncertainties.

The Documentary Sources

There are a number of medieval manuscripts or treatises that include technical descriptions of precious metal refining processes. Some, such as the *Mappae Clavicula* (Smith and Hawthorne 1974), are of limited value for interpreting archaeological finds, because the author's understanding of the processes he describes is often poor, and many of the recipes are just copied from earlier manuscripts. Others are far more useful, since most of the descriptions in them were obviously written by a practising metalworker, and the accuracy of their descriptions can be assessed by looking at those of well-understood processes. Quotations from two books will be used here to help explain the interpretation of archaeological finds. They have been chosen because they are well known and are readily accessible in good, modern English translations. They differ in their date, their approach, and their level of detail.

The first is Theophilus's *De diversis artibus*, which was written in Germany in the early 12th century, most probably by a Benedictine monk called Roger of Helmarshausen; the translation used here is that by Hawthorne and Smith (1979). Theophilus was writing for craftsmen like him who were making and decorating objects from metal they had bought or had been given. They were not necessarily knowledgeable about how these metals were obtained but had to be able to assess their purity as some of the techniques they used – for instance, fire gilding on copper alloys – would not work with impure metals.

The other is Agricola's *De re metallica*, written in Saxony in the mid-16th century by a doctor turned mining engineer, Georg Bauer. In their introduction, the translators, Hoover and Hoover (1950:iv), note: 'For 180 years it was not superseded as the textbook and guide to miners and metallurgists'. Agricola concentrates on how ores were located and on how metals were extracted from them, rather than on the fabrication of objects. However, assaying the richness of ores was an essential part of this work, as was separating alloys into their component metals. Because it was written over 400 years after *De diversis artibus*, it includes some techniques that were unknown to Theophilus, since they were developed only in the course of the Middle Ages.

Although neither book was written in or about Britain, archaeology has shown that medieval metalworking practices throughout Europe were similar, so comparing insular archaeology with central European texts is a valid approach – as the sections of this chapter demonstrate.

Cupellation

Variations of the cupellation process were used to separate silver from newly smelted lead, to test the purity of silver, and to purify precious metals that were being reused. Regardless of the specific purpose, cupellation is always based on the same principles, and its practice varied only slightly depending on the final intent (Bayley and Eckstein 1997).

Silver Production

Primary production of silver was mainly from lead-silver ores. Smelting these produced lead metal that contained variable but small amounts of silver. The precious metal could be separated by melting the metal in oxidising conditions, oxidising the lead, and removing it as litharge (PbO), and leaving a small amount of pure silver. In Britain, there is no definite physical evidence of medieval date for cupellation as a means of extracting silver from newly smelted lead, probably because the resulting litharge was all resmelted to recover the lead. A building used for lead-smelting in the 14th to 15th centuries at Derbyhaven on the Isle of Man has been suggested as a possible site of silver extraction (Megaw 1960–1961), although there is no real evidence to support this. Roman finds survive – for example, from Pentrehyling (Bayley and Eckstein 1998) – and similar material must have been produced in medieval times – for example, in the area of the 'mines of Carlisle' on the Cumberland/Durham/Northumberland/Yorkshire border, where there was a silver bonanza early in the 12th century (Blanchard 1992).

Refining Debased Silver

In contrast, there is plenty of medieval evidence of cupellation as a means of refining debased silver. Typical finds are described as 'litharge cakes' and have been found in contexts dating from the 8th to 13th centuries; Bayley and Eckstein (1997:109) provide a list of sites in England where litharge cakes had been identified up to 1995 and more recently have presented further details of this version of the cupellation process (Bayley and Eckstein 2006).

One site with an unusually large quantity of litharge cakes was in Thetford, where a late-12th/13th-century pit contained 20 kg of fragments (Bayley 1999). They are parts of roughly circular, concavo-convex 'cakes', which typically have diameters of around 150 mm (Figure 7.1). Many had a central depression in the upper, concave side, about 50–80 mm in diameter. Most of the cakes were about 20 mm thick near the rim but only 10 mm thick in the centre. They were mid-grey in colour and were noticeably heavier (denser) than slags normally are. X-ray fluorescence (XRF) analysis detected large amounts of lead together with lesser amounts of copper and traces of silver, but the latter was found only around the edge of the central depression. Further analysis of a polished section through one cake using an energy dispersive X-ray analyser on a scanning electron microscope (SEM-EDS) also detected calcium and phosphorus in significant amounts – the remains of the hearth lining, made of bone ash that had absorbed the litharge.

Figure 7.1 • Fragment of a litharge cake from Thetford (© English Heritage).

This version of the cupellation process can be represented by the following equation:

$$(Ag + Cu) + Pb + O_2 \rightarrow (PbO + CuO) + Ag.$$

The debased silver is melted with an excess of lead under oxidising conditions. Thus, the lead is oxidised to litharge (PbO), which acts in two ways: it reacts with the base metals mixed with the silver, oxidising them, and also forms fusible compounds with these metal oxides, which do not inhibit the oxidation of the remaining base metals (Percy 1870:177–78). The silver does not react and so is left behind when all the base metals have been oxidised and removed. The 'litharge cakes' are the litharge-impregnated lining of the hearth in which the cupellation was carried out.

Theophilus gives a good description of the process:

> Sift some ashes, mix them with water, and take a fire-tested earthenware dish of such a size that you believe the silver which is to be refined can be melted in it without running over. Put ashes in it, thinly in the middle and thickly around the rim, and dry it in front of the fire. When it is dry, remove the coals a little distance from the [wall of the] forge and put the dish with its ashes in front of the wall beneath the [tuyère] hole, so that the blast from the bellows will blow onto it. Then replace the coals on top, and blow until the dish is red-hot. Then put the silver into it, add a little lead on top, heap charcoal over it, and melt it. You should have at hand a stick cut from a hedge and dried in the wind, with which you should carefully uncover the silver and clean away whatever impurity you see on it. Then put on it a brand, that is, a stick burnt in the fire, and blow gently, using a long stroke [of the bellows]. When you have removed the lead by this blowing, if you see that the silver is not yet pure, again add lead, place charcoal on it, and do as you did before. However, if you see the silver boiling and jumping out, know that tin or brass is mixed with it, and finely crush a small piece of glass and throw it on the silver. Then add lead, put charcoal on, and blow vigorously. Then inspect it as before, take away the impurities of glass and lead with the stick, put the brand on it, and do as before. Keep on doing this until it is pure. (Hawthorne and Smith 1979:96–97)

Hawthorne and Smith (1979:97) comment that 'this is the earliest good description of the very ancient art of cupellation [. . .] Unlike most metallurgical operations which are done in relatively impervious refractories, this is intentionally done on a porous bed made of ashes (the best are bone ash but Theophilus does not specify) to absorb the oxidation products by capillary action'.

Ercker (1574) provides an illustration of refining silver by cupellation (Figure 7.2). The right hand part of the illustration shows the ash-lined

Figure 7.2 • Refining silver by cupellation (after Ercker 1574).

hearth (B) just below the tuyère, through which air enters from the bellows (D). On the left, the process is in progress with a fire (C) burning, and the craftsman holding a tool (F) with which he manipulates the fire and scrapes the litharge off the silver. Not surprisingly, Agricola has no description of refining debased silver as it is not part of any primary extraction process.

Small-Scale Refining and Assaying

Silver Smaller-scale cupellation of silver was also carried out in small dishes, known as tests or cupels. There are 16th-century examples made of bone ash (Figure 7.3), similar to the one that Agricola illustrates (Figure 7.4). Only the upper part of the cupel survives archaeologically (compare the heights of the cupels in Figures 7.3 and 7.4), since the bone ash is friable except where it is held together by the absorbed litharge. Agricola also shows the assay furnaces in which the cupels were used (Hoover and Hoover 1950:223–24) and the muffles that separated them from the charcoal in the furnace (Hoover and Hoover 1950:228). The lack of archaeological finds of assay furnaces from Britain to compare with these illustrations is not surprising, since they are small, above-ground structures and so would not be expected to survive. Even Oberstockstall, in Austria, where a large quantity of laboratory equipment has been

Figure 7.3 • Bone ash cupels from Legge's Mount, Tower of London (© English Heritage).

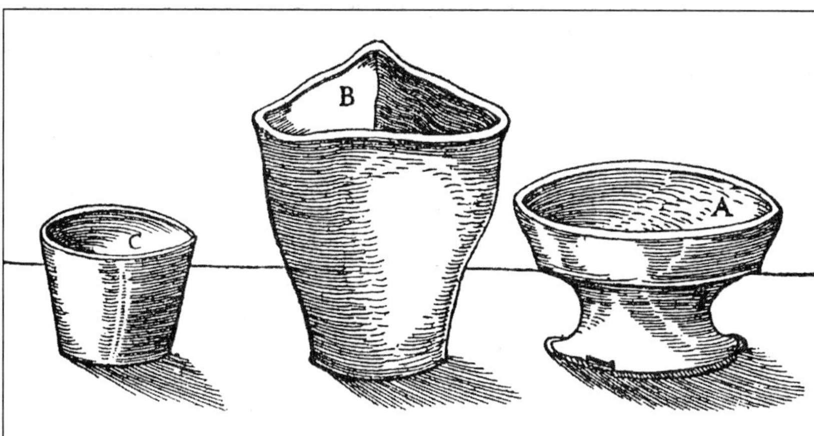

Figure 7.4 • Assaying vessels; the cupel (C) is to the left (after Agricola 1556).

found, produced only muffles (Martinón-Torres et al. 2003:Fig. 1) rather than complete furnaces.

Cupels could be used to purify a small quantity of silver but were probably mainly used to assay silver. The process is the same as on the larger scale described above but with the litharge being absorbed by the cupel. The metal sample was weighed before and after the process, the weight loss indicating how impure it was. Separation was normally very good, with no trace of silver detectable on the cupel. An exception is a

Figure 7.5 • Bone ash cupel from Cripplegate, London, with the refined silver *in situ* (© English Heritage).

16th/17th-century cupel from Cripplegate, London, where the whole mass of silver was found *in situ* (Figure 7.5 and Bayley 2003). The rough surface of the silver was caused by out-gassing of dissolved oxygen as the metal solidified, an indication of the purity of the metal (K. Eckstein, pers. comm.).

Archaeological finds suggest that 13th-century and earlier cupels were normally ceramic, either potsherds or purpose-made dishes (Figure 7.6). These did not give such good separation, because the litharge was not absorbed but reacted with the silica-rich ceramic fabric, forming a glassy layer in which some of the precious metal was mechanically trapped, so not all of it coalesced into a single mass (for example, Bayley 1992:Fig. 318 and Pl. LV). Analysis of these finds normally detects large quantities of lead and often enough silver to confirm that it was the metal being purified or assayed. Given the losses that are usually apparent, any assays using ceramic cupels can only have been approximate.

Gold For working with gold, the normal assay technique was known as touching. This was used to determine the fineness of gold in a given sample by rubbing it on a dark, fine-grained stone and comparing the streak left to those of a set of standards of known compositions. Agricola's description explains how it worked:

Figure 7.6 (a) • Ceramic cupel from Coppergate, York (Bayley 1992, no 2339) (© English Heritage).

Figure 7.6 (b) • Part of a potsherd reused as a cupel, from Coppergate, York (Bayley 1992, no 3961) (© English Heritage).

> Gold which contains silver, or silver which contains gold, is first rubbed on the touchstone. Then a needle in which there is a similar amount of gold or silver is rubbed on the same touchstone, and from the lines which are produced in this way, is perceived what portion of silver there is in the gold, or what portion of gold there is in the silver. (Hoover and Hoover 1950:253)

Touchstones are known from a number of early medieval sites (Moore and Oddy 1985). Some, including Southampton (M. Garner, pers. comm.) and Winchester (Oddy and Tylecote 1990), have visible gold traces, but others, such as the one from York (Bayley 1992:794), have no surviving evidence for their use.

There was, however, a far more basic but equally effective test. If gold contains significant amounts of base metals, its malleability is reduced, so just hitting it with a hammer can tell a skilled craftsman quite a bit about its purity. Theophilus describes what would be done:

> First test each piece of gold to see if it can be beaten with a hammer without cracking. Lay on one side whatever is not cracked, and on the other whatever is cracked, in order that it may be refined. (Hawthorne and Smith 1979:121)

This would of course leave no archaeological trace.

Ceramic cupels with gold droplets on them are known, and the fabrics are usually far more refractory than those used to refine silver (Figure 7.7). Lead is not normally present in significant amounts in the vitrified surface layer, which must therefore be produced by the reaction of ash from the fuel, or some deliberately added flux, with the cupel. Theophilus's description of refining impure gold by oxidising any base metals it contained could apply to these cupels:

> Then take a piece of burnt brick and cut a little groove in it big enough to hold the amount of gold that is to be refined. If you do not have brick, make a little groove with an iron tool in a square piece of sandstone and put it in the coals and blow. When it is red-hot, put the gold on it, throw charcoal on top, and blow for a very long time. Take it out and beat it with a hammer. If it does not break, that is enough; but if it does break, put it back again on another stone and keep on doing this until it will not break when it is hammered [. . .] When this is done, melt all the gold together and reduce it to a single mass. (Hawthorne and Smith 1979:121)

This description also fits the blocks of fused and vitrified quartz chips with a depression in the upper surface found on the Coppergate site in York (Figure 7.8), which were thought to have been used as cupels for gold refining (Bayley 1992:749–51, Fig. 319). Blocks of stone used in a similar

Figure 7.7 • Ceramic cupel from Coppergate, York, with gold droplets visible in the vitreous surface (Bayley 1992, no 3957) (© English Heritage).

Figure 7.8 • Block made of quartz chips from Coppergate, York, used for refining gold (Bayley 1992, no 3963) (© English Heritage).

way have been noted from Helgö in Sweden (Bayley 1981) and among the Viking Age metalworking finds from Dublin (Bayley 1989).

Parting

Agricola provides us with a good summary of the parting process, one that covers both salt parting and acid parting (a later medieval development):

> Gold is parted from silver, or likewise the latter from the former, whether it be mixed by nature or by art, by means of aqua valens [nitric acid, HNO_3], and by powders which consist of almost the same things as this aqua. (Hoover and Hoover 1950:439)

Salt Parting

All the archaeological finds relating to Roman and earlier medieval parting indicate a solid state process relying on common salt (NaCl) as the active ingredient. It is possible to use a mixture of saltpetre (KNO_3) and vitriol ($FeSO_4$) in a similar way, but as yet there is no archaeological evidence for this process. Agricola describes many recipes for the powdered 'cement' that reacted with the silver, removing it from the gold-silver alloy (Hoover and Hoover 1950:453–57). However, Theophilus has a single description that seems to fit with the archaeological evidence:

> break into tiny pieces a tile or piece of burnt and reddened furnaceclay and when it is powdered, divide it into two equal parts by weight and add to it a third part of salt of the same weight. It should then be lightly sprinkled with urine and mixed so that it does not stick together but is just moistened. (Hawthorne and Smith 1979:109)

Agricola also realises that a similar mixture will work:

> brick dust and salt alone, especially rock salt, are able to extract all the silver and copper from the gold and to attract it to themselves. (Hoover and Hoover 1950:457)

The mixed gold-silver had to be hammered out into thin sheets to provide a large surface area at which the reaction could take place. The parting vessels were then filled with alternate layers of metal and the cement, as both Theophilus and Agricola describe:

> The cement should be placed, alternately with the gold, in new and clean pots in which no water has ever been poured. In the bottom the cement is levelled with an iron implement, and afterward the gold granules or leaves are placed one against the other, so that they may touch it on all

> sides; then, again, a handful of the cement, or more if the pots are large, is thrown in and levelled with an iron implement; the granules and leaves are laid over this in the same manner, and this is repeated until the pot is filled. Then it is covered with a lid, and the place where they join is smeared over with artificial lute, and when this is dry the pots are placed in the furnace. (Hoover and Hoover 1950:455)

The description continues, saying the parting vessels should be heated gently for 24 hours, so the metal in them does not melt: '[. . .] it is ample to have the fire hot enough that the pots always remained red [. . .]' (Hoover and Hoover 1950:456).

The active constituent in the 'cement' is the salt (NaCl), which Hoover and Hoover (1950) say is decomposed at high temperatures in the presence of silica and alumina (the burnt clay) to give hydrochloric acid and, probably, free chlorine, which reacts with the silver, forming silver chloride (AgCl). The urine is acidic and would probably aid the decomposition of the salt. Moisture, presumably from the combustion of the fuel, is another unspecified but vital reactant. Silver chloride is volatile and so was removed from the metal:

$$(Au + Ag) + NaCl \rightarrow Ag\ Cl\uparrow + Au.$$

The lid was an essential part of the parting vessel, because it prevented the loss of volatile reactants, allowing the silver chloride to be absorbed by the burnt clay in the cement and, to a lesser extent, by the walls of the vessel used. The silver could be recovered by cupelling the cement afterward.

The only large group of medieval parting vessels known is from the Coppergate and Piccadilly sites in York (Figures 7.9 and 7.10; Bayley 1992:751–54), although other fragments are known from Winchester (Bayley and Barclay 1990:182), Annetwell Street, Carlisle (Bayley Forthcoming), and the site in Thetford that produced the litharge cake fragments mentioned above (Bayley 1999). Only the York vessels are reconstructable (Figure 7.10), but Roman examples suggest that any fairly small hand-made or wheel-thrown vessel could be used (Bayley 1991a). What helps identify parting vessel sherds is the unusual pinkish-purple colour of their inner surfaces. This colour is almost certainly a byproduct of heating the vessels in the presence of quantities of salt, which would have removed some of the iron from the clay fabric as ferric chloride. This compound is volatile and unstable at the temperatures at which parting was carried out and would have been oxidised to haematite and then deposited on the surfaces of the vessels. It is the combination of a somewhat bleached appearance of the fabric (owing to the loss of iron from the clay) combined with this unusual but characteristic pinkish-purple hue

Figure 7.9 • Parting vessel sherd from Coppergate, York, showing traces of the cement adhering (© English Heritage).

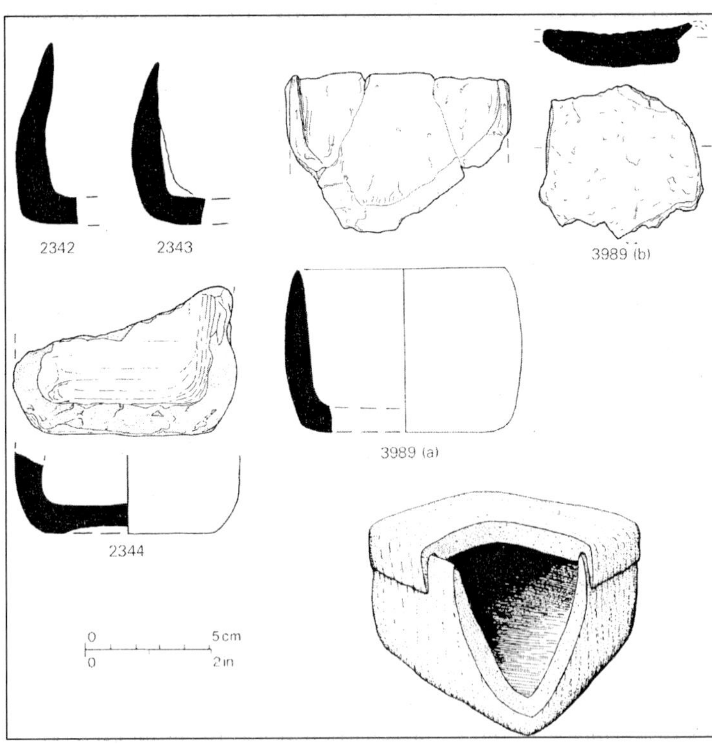

Figure 7.10 • Reconstruction of the parting vessels from York and the sherds on which it is based (after Bayley 1992, Fig. 320) (© York Archaeological Trust).

(owing to the presence of haematite) that visually identifies parting vessels. XRF analysis of their inner surface detects absorbed silver and occasionally traces of gold.

An alternative solid-state parting process involved the use of sulphur rather than salt as the main reagent. Theophilus describes this process (Hawthorne and Smith 1979:147), but no archaeological evidence for it has yet been recognised.

Acid Parting

In late medieval times distillation was introduced into western Europe (Taylor and Singer 1957); Theophilus does not mention it, but Ercker and Agricola do. In Britain, there are documentary references from the 14th century, but archaeological finds of definite distillation vessels are all from the 15th century or later (Tyson 2000:168ff). However, early applications seem mainly medicinal. There is no evidence for the production of strong mineral acids here before the 16th century, although Greenaway (1972) says they were known from the 14th century. Of these acids, it was nitric acid (*aqua valens*) that was used to part silver from gold by dissolving the silver while leaving the gold unaltered.

There are a number of shallow ceramic pots or dishes from Legge's Mount at the Tower of London that may have been used for this dissolution process. However, the archaeological finds that so far provide the best evidence for acid parting are those that demonstrate the production of nitric acid. From Legge's Mount and other sites in London there are many fragments of ceramic cucurbits, the vessel in which the reacting chemicals were heated. The acid that formed was then boiled off, condensed in the attached alembic, and collected in a receiving vessel. Agricola's illustration (Figure 7.11) shows the cucurbits almost completely enclosed in the furnaces, although the one in the single furnace to the right has a long neck, like the excavated examples from various sites in London (Figure 7.12). His recipes for making nitric acid would not all have worked (Hoover and Hoover 1950:439, note 3). Normally, to make nitric acid, saltpetre (KNO_3), vitriol ($FeSO_4$) or alum ($KAl(SO_4)_2$), and water were heated together. The reaction is

$$2KNO_3 + H_2O + FeSO_4 \rightarrow FeO + K_2SO_4 + 2HNO_3.$$

The archaeological finds of cucurbits are notable for the deposits of red iron oxides (Fe_2O_3) inside them. It is these deposits that identify the reaction, and hence the product.

Figure 7.11 • Distilling nitric acid (after Agricola 1556).

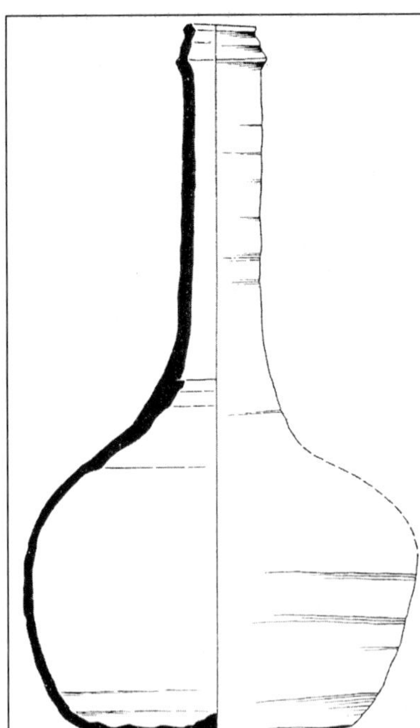

Figure 7.12 • Cucurbit from Legge's Mount, Tower of London; its height is ca 50 cm.

Discussion

The examples above have shown that in many cases the process in which archaeometallurgical finds were originally used can be deduced both from scientific analyses and from interpretation of, or just comparison with, descriptions in medieval treatises. In some cases, the identification can be made without reference to the texts, but in other cases, especially when a process had gone out of general use before the modern period (for example, salt parting), the texts suggest possibilities that can then be investigated by scientific analysis of the archaeological finds. As outlined above, the two main texts referred to here were written for different reasons and at different times, so for a particular process one is usually more illuminating than the other. Where they both deal with a process (for instance, salt parting), they usually provide complementary rather than identical information.

Problems with Using Medieval Texts

Even when using only texts such as those of Theophilus and Agricola, which are generally accepted as being technically competent, one can encounter problems. Some of these result from modern scholars' lack of familiarity with medieval Latin and their dependence on translations into modern languages. When the translators combine linguistic and metallurgical experience the problems are minimised, but the ambiguities and inconsistencies of the original Latin texts cannot be completely removed by translation, as the notes provided by Hawthorne and Smith (1979) and Hoover and Hoover (1950) demonstrate. Indeed, in the case of Theophilus, only copies of the original manuscript survive, and those are not all identical, so there is even uncertainty as to what should be translated!

Further problems derive from the fact that the texts contain many, sometimes contradictory, descriptions of precious metalworking and refining processes. Thus, when faced by archaeological objects whose use is unknown, one is tempted to select only those passages that appear to best illuminate the finds and describe the processes in which they were used. This approach has been followed to a large extent in the sections above. However, if the manuscripts are looked at as a whole, they suggest a wider range of processes. The question that must then be asked is why we have no archaeological analogues for these other processes. There are a number of reasons that may help account for these discrepancies:

- The process may not produce any durable products or byproducts; anything that was incorporated into archaeological deposits has long ago decomposed, corroded, or been washed away.

- Archaeologists may not have developed techniques to both recover and identify the diagnostic products or byproducts of a particular process.
- The process was an uncommon one, so it is not surprising that products or byproducts have not yet been identified. With a few honourable exceptions, it is only in the last 20 or 30 years that archaeological evidence for precious metal refining has been identified in Britain. It should be noted, however, that none of the manuscripts gives any idea of the frequency with which a particular process was used, especially when several alternatives are described.
- The descriptions in the manuscripts may be inaccurate, leading to problems in linking them to archaeological finds. This seems unlikely, because many of the descriptions can be seen to be accurate – for example, those used above.
- The descriptions may have been made deliberately confusing so knowledge would remain in the hands of the few. Again, this seems unlikely, since both Theophilus and Agricola were writing didactic texts, aimed at making knowledge more widely available.

Having dismissed many of the possible reasons, we are left with the partial nature of archaeological recovery as the most likely reason for the lack of one-to-one correlation between archaeological finds and medieval manuscripts. It is to be hoped that greater awareness and understanding among excavators and scientists will lead to the identification of some of the 'missing' processes in the future.

The use of medieval texts has a further set of inherent problems that are due to their authors' lack of modern chemical knowledge. This leads to recipes that contain a few ingredients that, from a modern perspective, appear superfluous. Their inclusion in the texts, from which we can infer their continued use, is best seen as a demonstration of the conservatism of medieval craftsmen and their disinclination to experiment. The other observation that can be made about the texts is their almost entire absence of theorising – as Bailey (1929) commented in the introduction to his translation of Pliny, 'in antiquity one was interested in facts or theories, but not, as a rule, in both'.

Conclusion

This brief review of medieval precious metal refining technologies has shown that a variety of processes was known to medieval craftsmen and was recorded by authors such as Theophilus and Agricola. Their texts help in interpreting archaeological finds, complementing the information that comes from scientific analyses of metalworking assemblages. Because there are far more archaeological finds than there are surviving texts, the contextual dating assigned to the finds provides us with a more detailed chronology for the introduction of new technologies and can also show when earlier technologies fell out of use. Examples of this are the introduction of acid parting and the disappearance of ceramic cupels.

Acknowledgments

I would like to thank Thilo Rehren for inviting me to present this paper and for subsequently encouraging me to prepare it for publication.

References

Agricola, G. 1556. *De Re Metallica Libri XII.* Basel: J. Froben and N. Episopius.

Bailey, K. C. 1929. *The Elder Pliny's Chapters on Chemical Subjects, Part I.* London: Edward Arnold.

Bayley, J. 1981. *Examination and Analysis of Some 'Crucibles' from Helgö.* Ancient Monuments Laboratory Report 3459.

———. 1989. *Metalworking Finds from Viking Dublin.* Ancient Monuments Laboratory Report 76/1989.

———. 1991a. 'Archaeological evidence for parting', in E. Pernika and G. A. Wagner (eds.), *Archaeometry '90.* Basel: Birkhäuser Verlag, 19–28.

———. 1991b. 'Processes in precious metal working', in P. Budd, B. Chapman, C. Jackson, R. Janaway, and B. Ottaway (eds.), *Archaeological Sciences '89.* Oxford: Oxbow Books, 125–31.

———. 1992. *Non-Ferrous Metalworking at 16–22 Coppergate*, The Archaeology of York 17/7. London: CBA.

———. 1993. 'Precious metal working in Anglo-Saxon England', in C. Eluère (ed.), *Outils et ateliers d'orfèvres des temps anciens.* Paris: Antiquités Nationales mémoire 2, 137–40.

———. 1999. 'Metalworking evidence [from excavations at Guildhall Street, 1989, Site 25296]', in P. Andrews and K. Penn (eds.), *Excavations in Thetford, North of the River, 1989–1990.* East Anglian Archaeology 87. Gressenhall: Norfolk Museums Service.

———. 2003. *Metalworking Debris from Cripplegate Buildings (WFG 18), City of London.* Centre for Archaeology Report 80/2003.

Bayley, J. Forthcoming. In I. Caruana (ed.), *The Roman Forts at Carlisle: Excavations at Annetwell Street 1973–1984.* Britannia Monograph. London: Society for the Promotion of Roman Studies.

Bayley, J., and K. Barclay. 1990. 'The crucibles, heating trays, parting sherds, and related material', in M. Biddle (ed.), *Object and Economy in Medieval Winchester.* Winchester Studies 7.ii. Oxford: Oxford University Press, 175–97.

Bayley, J., and K. Eckstein. 1997. 'Silver refining: Production, recycling, assaying', in A. Sinclair, E. Slater, and J. Gowlett (eds.), *Archaeological Sciences 1995.* Oxford: Oxbow Monograph 64, 107–11.

———. 1998. *Metalworking Debris from Pentrehyling Fort, Brompton, Shropshire.* Ancient Monuments Laboratory Report 13/1998.

———. 2006. 'Roman and medieval litharge cakes: Structure and composition', in J. Pérez-Arantegui (ed.), *Proceedings of the 34th International Symposium on*

Archaeometry, Zaragoza, 2004. Zaragoza: Institución Fernando El Católico, 145–53.

Blanchard, I. 1992. 'Technical implications of the transition from silver to lead smelting in twelfth-century England', in L. Willies and D. Cranstone (eds.), *Boles and Smeltmills.* London: Historical Metallurgy Society, 9–11.

Ercker, L. 1574. *Beschreibung allerfürnemisten mineralischen Ertzt und Berckwercksarten.* Prague.

Greenaway, F. 1972. Introduction, in S. Moorhouse (ed.), 'Medieval distilling-apparatus of glass and pottery', *Medieval Archaeology* 16:79–121.

Hawthorne, J. G., and C. S. Smith. 1979. *Theophilus: On Divers Arts.* New York: Dover.

Hoover, H. C., and L. H. Hoover. 1950. *Georgius Agricola: De re metallica.* New York: Dover.

Martinón-Torres, M., Th. Rehren, and S. von Osten. 2003. 'A 16th-century lab in a 21st-century lab: Archaeometric study of the laboratory equipment from Oberstockstall (Kirchberg am Wagram, Austria)', *Antiquity* 77(298), Project Gallery [http://antiquity.ac.uk/ProjGall/martinon/].

Megaw, B. 1960–1961. '"Backenaldwath" and the medieval lead mines', *Journal of the Manx Museum* 6(77):105–07.

Moore, D. T., and W. A. Oddy. 1985. 'Touchstones: Some aspects of their nomenclature, petrography and provenance', *Journal of Archaeological Science* 12:59–80.

Oddy, W. A., and R. F. Tylecote. 1990. 'Objects associated with gold working', in M. Biddle (ed.), *Object and Economy in Medieval Winchester.* Winchester Studies 7.ii. Oxford: Oxford University Press, 76–78.

Percy, J. 1870. *Metallurgy* (Vol. 3). London: John Murray.

Roesdahl, E. 1977. *Fyrkat: En jysk vikingeborg. II: Oldsagerne og gravpladsen.* Copenhagen: Herm H J Lynge.

Smith, C. S., and J. G. Hawthorne. 1974. '*Mappae Clavicula*: A little key to the world of medieval techniques', *Transactions of the American Philosophical Society* (new series) 64(4).

Taylor, F. S., and C. Singer. 1957. 'Pre-scientific industrial chemistry', in C. Singer, E. J. Holmyard, A. R. Hall, and T. I. Williams (eds.), *A History of Technology.* (Vol. II). Oxford: Clarendon Press, 347–74.

Tyson, R. 2000. *Medieval Glass Vessels Found in England c AD 1200–1500.* York: CBA Research Report 121.

CHAPTER 8

Lustre Recipes for Hispano-Moresque Ceramic Decoration in Muel (Aragón, Spain), or 'How Much a Little Copper Weighs'

JOSEFINA PÉREZ-ARANTEGUI AND CARLOS PARDOS

ABSTRACT *Lustre pottery is a renowned ceramic decoration obtained by a complex technical process. It was first produced in the 9th century* C.E. *in the Middle East. The manufacture of lustre was seldom described in detail before the late 19th century; recipes were kept zealously secret by the guilds. A few documents, such as letters or travel books, collected by curious nonspecialists surprised by the appearance of lustre pottery, are the only historical sources relating to their manufacturing techniques. This chapter outlines the technical sequence of production of lustre ceramics as identified through scientific analyses of archaeological remains. A specific 16th-century recipe recorded in Muel is then presented and compared to archaeological finds from the same area. The analytical approach helps to interpret the vague directions given in the description, and it reveals that, most likely, the informant was deliberately hiding a crucial part of his recipe in order to hinder competition.*

A Ceramic Decoration: Reduced-Pigment Lustre

Lustre pottery is an exquisite ceramic decoration obtained by a complex technical process (Pérez-Arantegui et al. 2001; Pradell et al. 2004) (Figure 8.1). It was first produced in the 9th century C.E. in the Middle East (Iraq). Once the shaped ceramic object has dried, it is prefired ('biscuited') at high temperature, usually 900–1000°C, and then glazed and fired again. The decoration is then painted on this fired glaze with

Figure 8.1 • Lustre pottery from Muel (Zaragoza), 16th century C.E. (photo J. Garrido; © Museo de Zaragoza, Spain).

a mixture of mineral pigments dissolved in vinegar. The ceramic is then subjected to a third firing at a low temperature in a reducing atmosphere. When the object is cool and taken out of the kiln, the decoration appears covered with a black film, but it finally shows a beautiful metallic lustre when the object is washed and cleaned by rubbing.

Among the different ceramic techniques often termed as 'lustre', the decorative tradition, on which this study is focused, is the so-called 'reduced-pigment lustre'. Almost all the lustre made before 1800 belongs to this family, and it includes all Islamic, Spanish, and Italian lustre ware. However, there are three other decoration techniques also often called 'lustre', and it is desirable to distinguish them: 'resinate lustre' (also called 'liquid gold' or 'liquid platinum', produced in oxidising atmosphere), 'gilding' (the application of gold leaf or powdered gold to ceramics with adhesives) and 'reduced-glaze lustre' (all the glaze is reduced and lustre usually covers the entire surface of an object) (Caiger-Smith 1985). The first two techniques result in a metallic film decoration, radically different from the last one.

The 'reduced-pigment lustre' technique spread from the Middle East to Muslim Spain, where lustre pottery production was developed for several centuries, and it even spread to other European countries, such as Italy. Although some Spanish centres, such as Málaga, Manises, Paterna, and Valencia, are renowned because of their lustre ceramics, this chapter focuses on another inland area producing this type of ceramic decoration in Spain: Muel, a small village near Zaragoza (Alvaro Zamora

2002; Rose-Albrecht 2002). Lustre pottery was produced in Muel for a long period but especially during the 16th century, before the expulsion of the *Moriscos* ordered by Philip III in 1610. Most of the people living in Muel at that time were Muslim, although they had been obligatorily baptised since 1522 and then called *Moriscos* or 'Muslim converts to Christianity'. They made fine tin-glazed objects decorated with golden or yellow designs, similar to those used on metallic vessels. After 1610, lustre production continued in Muel until the mid-17th century, probably by potters coming from Catalonia or Valencia.

Lustre ceramics were very valuable objects, real luxury products and very highly admired. Although the overall production technique is more or less known in general terms, the ingredients mixed in the ancient recipes used to produce lustre are much more difficult to specify. The production of lustre was seldom described in detail before the late 19th century C.E., which is expectable if we consider that these objects had a very complicated manufacture and that their technology was, for centuries, kept zealously secret by potters, who were not interested in its diffusion. Only a few documents mention characteristics of pigments or procedures of lustre technology before 1800: these are treatises, letters, and journey descriptions, collected by curious or nonspecialist individuals intrigued by the appearance of lustre pottery. Both reasons, the secrecy and the indirect nature of the information, make us question the accuracy of some historical descriptions. Moreover, no written records exist for the earliest centuries of lustre production; therefore, there is a risk of interpreting the ancient technology in terms of current techniques. Another difficulty in understanding ancient recipes is the terminology used to describe materials and ingredients, given the long period and the large geographical area considered.

Ancient Lustre Recipes

In the following paragraphs, some recipes for manufacturing lustre are summarised (see Tables 8.1 and 8.2), following primarily the references collected by A. Caiger-Smith (1985). Special attention is paid to mixtures used to produce lustre, particularly ingredients such as copper, silver, and gold, that could be detected in the analysis of this decoration in archaeological examples.

The earliest published reference is the *Jowhar-name-ye Nezami* (Porter 2003), a treatise on precious stones, minerals, and metals written by Muhammad al-Jowhar al-Neyshapuri in 1196 C.E. in the area that is now Iran. The text is divided in four chapters, the last one devoted to glazes and metallic lustre. In this fourth chapter, 24 recipes explaining lustre with different colours are described, including white, black, yellow,

Table 8.1 Summary of some lustre recipes described in the *Jowhar-name-ye Nezami* (Porter 2003).

RECIPE	1 (Red)	2 (Red)	3 (Red)	4 (Golden)	5 (Golden)	6 (Golden)	8 (like fire)	22 (like fire)
Manganese	Manganese (MnO_2?)	Manganese (MnO_2?)				Manganese (MnO_2?)		
Iron	Yellow vitriol ($Fe_2(SO_4)_3$?)	Gold marcasite (FeS_2)		Qolqotar (Fe_2O_3 or $Fe_2(SO_4)_3$?)	Gold marcasite (FeS_2)			
Sulphur	White sulphur ($CaSO_4$?)	White sulphur ($CaSO_4$?)				Yellow sulphur (S)		Yellow sulphur (S)
		Cinnabar (HgS)	Cinnabar (HgS)	Cinnabar (HgS)		Cinnabar (HgS)	Cinnabar (HgS)	
			Realgar (As_2S_2)			Realgar (As_2S_2)		Realgar (As_2S_2)
						Orpiment (As_2S_3)	Orpiment (As_2S_3)	

		Vitriol			Blue vitriol (qalqand)	Qalqandis (white alum? $KAl(SO_4)_2.12H_2O$) Vitriol Lapis stone (cobalt compound?)	Green vitriol	Vitriol
Silver	Silver burned with sulphur (Ag_2S)	Roasted silver (Ag_2O)	Roasted silver (Ag_2O)	Silver burned with sulphur (Ag_2S)	Roasted silver (Ag_2O) Silver scoria	Roasted silver (Ag_2O)	Roasted silver (Ag_2O)	Silver
Copper			Verdigris (Cu $(CH_3COO)_2.H_2O$)		Verdigris (Cu $(CH_3COO)_2.H_2O$)	Verdigris ($Cu(CH_3COO)_2.H_2O$) Roasted copper Roasted brass	Verdigris (Cu $(CH_3COO)_2.H_2O$)	Red copper

Table 8.2 Summary of some other lustre recipes described in the text.

RECIPE	Abu'l Qasim (1301)	Di Baldassare (14th c AD)	Reyner (1514–1519)	Piccolpasso (1557)	Cock (1585)	Causada (1765)	Muel potters (1765)	Manises potters (1785)
Iron	Yellow vitriol ($Fe_2(SO_4)_3$?) Gold marcasite (FeS_2) Silver marcasite ($FeAsS$)							
			Red ochre (Fe_2O_3)	Armenian bole (Fe_2O_3)	*Almagre* (Fe_2O_3)	*Almagre* (Fe_2O_3)		Red ochre (Fe_2O_3)
Sulphur			Sulphur (S)					Sulphur (S)
			Cinnabar (HgS)	Cinnabar (HgS)	Vermilion (HgS)	Vermilion (HgS)		
	Realgar (As_2S_2) Orpiment (As_2S_3)	Orpiment (As_2S_3)						
					'*Alumbre*'(?) ($KAl(SO_4)_2 \cdot 12H_2O$)			
		Cofollo (PbS)						

Silver	Silver burned with sulphur (Ag_2S)	Silver	Silver coins	Burned silver (Ag_2O)	Silver coins	Silver burned with sulphur (Ag_2S)	Silver (or gold leaf)	Silver burned with sulphur (Ag_2S)
Copper	Roasted copper				'*Alambre*' (Cu)			
			Faret (CuS)	*Ferretto* (CuS)		Copper burned with sulphur (CuS)		Copper burned with sulphur (CuS)
Others							Litharge (PbO) and minium (Pb_3O_4)	

golden, silver, red, blue, and green. Porter (2003) translated the 24 recipes into French, although he discusses only the first four. According to all these recipes, potters basically used manganese oxides, iron sulphide or sulphate, cinnabar (mercury sulphide, HgS), sulphur, silver sulphide, and copper compounds. The different ingredients are summarised in Table 8.1, where we have included only seven recipes that can produce a truly metallic decoration (in some cases, recipes are not complete, or they would probably produce a glaze and not a lustre), with similar colours to Spanish lustre. In all cases, ingredients were mixed in vinegar.

In 1301 C.E., Abu'l Qasim of Kashan wrote a treatise on ceramics in which one section is devoted to metallic lustre. The manuscript was translated into German in 1935 and into English in 1973, and then published and annotated by J. W. Allan (1973). This was the earliest known written description of lustre manufacture until a few years ago, and it has been one of the most frequently cited in subsequent publications.

Abu'l Qasim describes the following ingredients to make lustre (Allan 1973): 1½ *mans* (1 *man* = 3 kg) of red and yellow arsenic (realgar, As_2S_2 and orpiment, As_2S_3), 1 *man* of gold and silver marcasite (FeS_2; FeAsS), ½ *man* of yellow vitriol (probably iron sulphate or copper pyrites [Allan 1973:117]), and ¼ *man* of roasted copper; all of them mixed to a paste and ground. After these elements have been mixed, a quarter of this paste is mixed with 6 *dirhams* (1 *dirham* = 11.5 grams) of pure silver, which has been previously burned and ground with sulphur, and the new mixture is ground again on a stone for 24 hours until it is extremely fine. This mixture is dissolved in some grape juice or vinegar. Also, certain compounds such as lead oxide and copper acetate are added in some cases. The main recipe is similar to recipes number 5 and 22 collected by Muhammad al-Jowhar al-Neyshapuri (Porter 2003) (see Tables 8.1 and 8.2), and there are also some ingredients in common with other descriptions.

Another early description of lustre making was published by Gaetano Milanesi in 1864 as a recipe 'for the colour which looks like gold'. In this recipe, attributed to the Florentine Benedetto di Baldassare Ubriarchi (born in 1377 C.E.), the main compounds were *cofollo* (powdered galena, lead sulphide, PbS), orpiment (arsenic sulphide, As_2S_3), and a little fine silver (Milanesi 1864).

These three first cases are the earliest descriptions known to us. Since the 16th century, several recipes from Italy and Spain have been also collected. In this chapter, we draw especially from Spanish sources.

In the *Libre de les Fornades* (1514–1519 C.E.) of Nicolau Reyner of Barcelona (Ainaud de Lasarte 1942:94), a recipe for golden lustre is described as 3 ounces of cinnabar, 3 ounces of *faret* (copper sulphide), 12 ounces of red ochre (red earth with a high content of iron oxides), 4 bars of sulphur, and 2 Barcelona *reales* of 3.5 g (silver coins). This recipe is very similar to that described by Piccolpasso in Italy in 1557 C.E., where the mixture included red earth, Armenian bole (a clay type containing a high amount of ferric oxide), *ferretto* from Spain (interpreted as copper sulphide), cinnabar (mercury sulphide, HgS), and burnt silver (Piccolpasso 1976).

Of special interest for this study is the recipe collected by Henrique Cock. Notary and member of the royal bodyguard of Philip II (Morel Fatio and Rodriguez Villa 1876:30), he described the technique of lustre produced in Muel (Zaragoza, Spain), just as *Morisco* potters told him during a journey to Zaragoza in 1585. They used a mixture of strong vinegar with 2 silver *reales* (small coins) previously powdered, vermilion (mercuric sulphide, HgS), *almagre* (iron oxides), and a little '*alambre*'. This last ingredient is identified by some authors as alum (*alumbre* in Spanish), that is, aluminium potassium sulphate (Caiger-Smith 1985:215). However, in other Spanish references (Alvaro Zamora 2002:104), it was identified as copper (ancient meaning of *alambre*, used for naming pieces of copper and copper alloys). Both interpretations of this ingredient could be possible, given that both copper and sulphate are elements necessary for lustre formation (Pradell et al. 2004), but it would be necessary to check the original Spanish manuscript (in the Bibliothèque Nationale de Paris) to verify the original spelling, and even so, we would not be sure if Cock's interpretation was correct, owing to the phonetic similarity of both words, since he was not a specialist in ceramics.

Two other Spanish recipes were written in 1765. The first one came from Alcora (Castellón, Spain), and it belonged to Jacinto Causada, a potter of Aragonese origin, who used copper and silver sulphides, mixed with vermilion (mercuric sulphide, HgS) and *almagre* (iron oxides), to which he also added the dirt or remains (called *cosela*) of lustre pigment, collected when objects were rubbed and washed after lustre firing (Escrivá de Romaní 1945:194). Copper and silver sulphides were obtained by placing the metals in a big pot between layers of sulphur grains and heating them in a fire until all the sulphur was burned. The product obtained was subsequently ground.

The second recipe was from Muel potters working in 1765, it was included in the *Ordinaciones de la Cofradía y Gremio de Alfareros y Vajilleros* (*Ordinances of the Fraternity and Guild of Potters*), a document that has disappeared at present, and its ingredients are 2 parts litharge (PbO), 1 part minium (Pb_3O_4), and either ½ part powdered silver, or, in its place, powdered gold leaf (Galiay Sarañana 1947:18).

The last recipe quoted here is a description included in the report written in 1785 at the request of the chief magistrate of Valencia, to be sent to the Count of Floridablanca (Riaño 1878). It is a lustre-making recipe from Manises (Valencia, Spain). According to this source, potters obtained lustre by mixing copper (3 ounces), silver (1 *peseta*, Spanish coin), sulphur (3 ounces), red ochre (12 ounces), and strong vinegar (1 *azumbre*, about 2 litres). Three pounds (in this case, 1 pound = 12 ounces = 360 grams) of scoria or dirt remaining from lustre firing were also added to the mixture. The pigment preparation described is similar to that from Alcora in 1765: layers of powdered sulphur are placed in an earthen cooking pot together with small pieces of copper, and in between them, the silver coin; arranged in this manner, the cooking pot is placed on a fire and is heated until the sulphur is consumed; next it is taken out of the pot, and when the product is cold, it is ground up to a very fine powder; then the ochre and the scoria are mixed in and stirred well by hand, before grinding the mixture again until it becomes powdery. This powder is still fired for six hours, and afterward it is ground once again together with vinegar, finally rendering it ready for painting.

A summary of more recent recipes (19th–20th centuries) used to produced lustre following ancient traditions in Manises (Valencia, Spain), is given by Pérez Camps (1998). In general, these recipes bear great similarities to the 1785 recipe (Riaño 1878), except for the new commercial form of some ingredients such as copper oxide or silver nitrate.

Elements in the Composition of Lustre from Muel

The workshops in Muel produced lustre pottery with very different colours: golden-yellow, red, reddish-brown, light-brown. Given the variety of colours as well as the appearance of the glossy lustre – some of them with a real metallic lustre, some others not – the question emerges of how these decorations were achieved. The full answer to this question is very complex, and it has given rise to a large project in order to reproduce the style and the way of producing ancient lustre. This chapter concentrates on a comparison between what written sources said that potters used to produce lustre and what can be revealed by modern analyses of ancient samples.

In the case of Muel, no archaeological evidence of the pigments used to make lustre has been found, contrary to what happened in other ceramic workshops, where the lustre recipe could be identified to consist of clay, iron oxides, copper and silver compounds, and cinnabar (Molera et al. 2001). Therefore, only finished objects decorated with lustre coming from museums or archaeological sites could be analysed.

The fragments investigated date probably from the 16th century to 1610 and belong to the collection of the Museum of Zaragoza. The sample chronology was chosen to better attain the project aim of reconstructing the Hispano-Moresque style and the way of producing ancient lustre in Aragón.

Chemical composition of the lustres was studied on the surface of lustre ceramics by an energy-dispersive X-ray (EDS) analysis system attached to a scanning electron microscope (SEM). Owing to the structure (quasi-spherical metallic nanoparticles embedded in an amorphous matrix, see Figure 8.2d) and the extreme thinness of the lustre (less than 1 micron) (Molera et al. 2002; Pérez-Arantegui and Larrea 2003; Pérez-Arantegui et al. 2001, 2004), and given the analytical technique used, the chemical composition obtained included elements responsible for the lustre decoration (copper and silver) but also elements forming the underlying tin-glaze (silicon, lead, potassium, calcium, aluminium, and tin) in which the lustre is embedded (Figure 8.2).

As a result of the preceding problems, only two of the main elements possibly included in the potter's recipe could be determined and studied: copper and silver. Because of the nature of the lustre formation process,

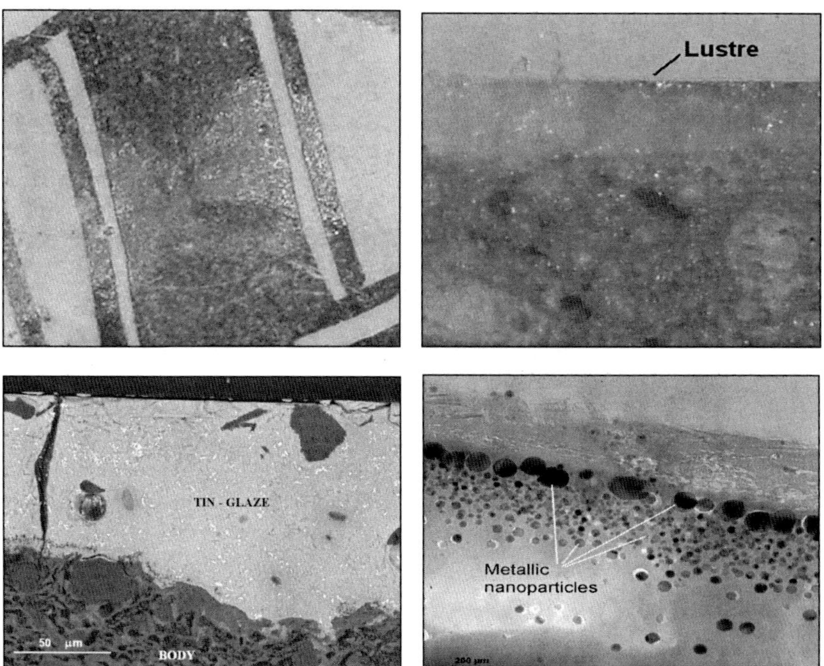

Figure 8.2 • (*top left*) Image of a lustre surface. (*top right*) Micrograph of a cross-section of lustre pottery; lustre decoration is just on the surface. (*bottom left*) Backscattered-electron image of a cross-section of lustre pottery. (*bottom right*) Transmission-electron image of a cross-section of lustre pottery; see lustre dimensions (less than 0.05 µm) compared to tin-glaze dimensions (about 100 µm).

only copper and silver remain within the glaze producing the lustre. The rest of the ingredients take part in the process but remain in the scoria, which is rubbed and washed off after firing (Molera et al. 2005; Pradell et al. 2004). Some data of Cu/Ag weight ratios are shown in Table 8.3 for different lustre samples. Owing to the small thickness of the lustre, we think that it is more interesting to consider the proportion of both elements in the possible recipes, rather than the absolute values.

Analytical results show that it is possible to differentiate some lustres with a relatively high proportion of silver (samples ZA03, ZA05, and ZA09) – with one case (ZA09) where only silver is detected. These lustres show golden and golden-brown colours. In a second group, brown and red-brown coloured samples show a Cu/Ag ratio between 1 and 2, with some analytical spots where only copper is detected. Finally, there are examples with a Cu/Ag ratio higher than 2, or where only copper was detected; in some copper-red samples (ZA11, ZA12, and ZA14) the concentration of copper is even higher. Therefore, it is clear that several mixtures of copper and silver were prepared and applied to produce lustre decorations of different colours.

Table 8.3 Cu/Ag ratio (in weight) in different lustre samples (*Ag: only detected silver; *Cu: only detected copper); three determinations were done on each sample scanning an area of 1 mm² at a magnification 100x.

Cu/Ag	1	2	3	Lustre colour
ZA09	0.52	*Ag	*Ag	Golden (olive-oil)
ZA03	0.83	0.47	0.65	Golden-brown
ZA05	0.88	0.87	0.70	Golden-brown
ZA08	*Cu	0.54	1.52	Golden-brown
ZA07	*Cu	1.03	*Cu	Golden-brown
ZA20	0.86	1.15	0.89	Brown
ZA17	*Cu	0.90	1.01	Brown
ZA10	1.51	1.77	1.47	Brown
ZA18	1.00	1.80	1.91	Red-brown
ZA15	1.63	*Cu	1.45	Red-brown
ZA21	1.51	2.01	1.64	Red-brown
ZA19	1.78	2.27	1.22	Red-brown
ZA22	1.87	*Cu	*Cu	Brown
ZA02	*Cu	1.95	*Cu	Brown
ZA16	2.46	4.40	2.03	Brown
ZA23	1.48	2.08	3.39	Brown
ZA24	*Cu	3.07	1.57	Red-brown
ZA11	*Cu	*Cu	2.75	Copper-red
ZA14	1.42	*Cu	*Cu	Copper-red
ZA12	3.13	3.67	3.10	Copper-red

Mercury cannot be detected in finished lustre because it evaporates during firing, but its role in lustre production (Pradell et al. 2004) and its presence (as cinnabar, HgS) in other Spanish medieval workshops have been verified (Molera et al. 2001).

Coming back to the relationship between these elements and the ancient recipes, we could consider two of the texts summarised above, since both of them were collected in Muel: one is the recipe described by Henrique Cock (Morel Fatio and Rodriguez Villa 1876:30), and the other one is that recorded in the 1765 *Ordinaciones* (Galiay Sarañana 1947:18). However, it becomes obvious that the former appears more relevant.

Starting with the 18th-century recipe: this seems to describe the ingredients of a metallic film decoration obtained by 'gilding' or 'liquid gold' ('resinate lustre') techniques, rather than the 'reduced-pigment lustre' produced in the village and documented in the archaeological samples. According to the text, silver or gold was introduced already reduced, and a high quantity of lead (litharge and minium) helped the formation of a transpartent glaze on the ceramic surface, where silver or gold would be fixed like a film, that is, 'gilding'. Alternatively, dissolved metals, suspended with fluxes in a resin-balsam, were fired in an oxidising atmosphere and reduced by the resin-balsam burning, that is, 'resinate lustre'.

In the traditional 'reduced-pigment lustre' technique, if gold was included in the glaze and reduced during the lustre process, very tiny particles of metallic gold were present within the glaze, and the final lustre colour was ruby red (or blood red), and not golden. In the Muel objects analysed, red lustre is always due to the presence of copper, and gold was never found in the analyses. Probably, Galiay Sarañana (1947:18) included this 18th-century recipe from Muel in his book because the ancient document (1765) described it as a golden decoration or metallic lustre, but in practice this corresponded to other lustre techniques, already known and popular in Europe from about 1700, when European fashion developed an interest in tableware with metallic surfaces. These are, however, very different in style and sheen from the traditional 'reduced-pigment lustre' ceramics (see Caiger-Smith 1985, Chapter 9). The fact that gold was never found in analyses of traditional 'reduced-pigment lustre' samples from Muel, in spite of some golden colours, proves that we cannot extrapolate any written description to explain archaeological remains. It is also noteworthy that general descriptive terms (such as *golden*, *metallic*, and even *lustre*) may not be precise enough for archaeological purposes, even if the visual aspect of an 18th-century porcelain with golden gilding decoration is completely different from the iridescent golden lustre on a 16th-century Muel ceramic.

Coming now to the other written source: Henrique Cock's recipe has been interpreted in two different ways, stemming from different readings of the term *alambre*. If Caiger-Smith's interpretation is considered, lustre ingredients would be powdered silver (2 *reales*), mercuric sulphide, a little alum (aluminium-potassium sulphate), and vinegar (Caiger-Smith 1985:215). In this case, the ancient recipe would agree only with the first Cu/Ag ratios, as shown in Table 8.3, that is, the highest silver proportions or where only silver was detected (samples ZA03, ZA05, or ZA09). Even though all analyses showed the presence of copper in the lustre, this could be explained by a small amount of copper introduced as a part of a silver alloy (coins never were pure silver). This Cu/Ag proportion would produce the most golden lustre decorations, with colours similar to olive oil.

If Alvaro Zamora's interpretation is followed, Muel lustre was made with powdered silver (2 *reales*), mercuric sulphide, 'a little copper', and vinegar (Alvaro Zamora 2002:104). This possibility could agree with several of the Cu/Ag ratios reported, between 1 and 2 or even higher than 2 – the question would be *how much a little copper weighs*. If the nonspecified *real* was similar to the Barcelona *real* described by Ainaud de Lasarte (1942:94), which weighed 3.5 g, then the quantity of silver added would be approximately between 6 and 7 g; thus 'a little copper' could be, for example, 2 g (Cu/Ag = 0.3), or 5 g (Cu/Ag = 0.8), or 10 g (Cu/Ag = 1.5), or 20 g (Cu/Ag = 3), and so on. These combinations would produce a wide range of possibilities and colours, and this is exactly the secret that the potters did not tell Henrique Cock.

Conclusion

Historical documents describe characteristics of pigments and procedures of lustre technology before 1800, but the potters' secrecy and the indirect nature of the information made us question the accuracy of some of the collected descriptions. In addition, there is a risk of interpreting ancient technologies in terms of current techniques, and this is why an extensive analytical study of lustre pottery was necessary in order to identify the pigments used and to understand the ways of production. Analytical results confirm that some ingredients (copper and silver) were always used in the lustre ceramics studied, and in a wide range of proportions, leading to the various colours achieved. On the contrary, the suggested use of other elements such as gold, is refuted by the chemical analyses of ancient 'reduced-pigment lustres' from Muel. Altogether, this study shows that analytical and textual information, if critically compared, provide a richer picture of past technologies.

Acknowledgments

This study was supported by 02/COOPERA/ES1368 research project in the INNOVARAGON programme (European Union [FEDER], Diputación General de Aragón and Instituto Tecnológico de Aragón, Spain) and by a 2004–2005 research project of the Comunidad de Trabajo de los Pirineos. We also thank Dr Miguel Beltrán and Dr Juan Paz (Museo de Zaragoza) for allowing us to study the ceramic material.

References

Ainaud de Lasarte, J. 1942. 'La loza dorada y alfarería barcelonesa, siglos XV–XVI', *Anales y boletín de los museos de arte de Barcelona* 1:89–104.

Allan, J. W. 1973. 'Abu'l-Qasim's treatise on ceramics', *Iran* IX:111–20.

Alvaro Zamora, I. 2002. *Cerámica aragonesa* (Vols. 1–3). Zaragoza: Ibercaja.

Caiger-Smith, A. 1985. *Lustre Pottery: Technique, Tradition and Innovation in Islam and the Western World.* London: Faber & Faber.

Escrivá de Romaní, M. 1945. *Historia de la cerámica de Alcora.* Madrid: CSIC, Instituto Diego Velázquez.

Galiay Sarañana, J. 1947. *Cerámica aragonesa de reflejo metálico.* Zaragoza: Institución 'Fernando el Católico'.

Milanesi, G. 1864. 'A fare le invetriatura delle scodelle do maiolica', in *Dell'Arte del vetro per musaico tre trattatelli dei secoli XIV e XV ora per le prima volta pubblicatti.* Bologna: Romagnoli.

Molera, J., M. Mesquida, J. Pérez-Arantegui, T. Pradell, and M. Vendrell. 2001. 'Lustre recipes from a medieval workshop of Paterna', *Archaeometry* 43(4):455–60.

Molera, J., J. Pérez-Arantegui, T. Pradell, and M. Vendrell-Saz. 2002. 'La céramique musulmane et mudéjare à reflets métalliques: Une approche technique', in *Le calife, le prince et le potier: Les faïences à reflets métalliques.* Paris: Réunion des Musées nationaux, 214–19.

Molera, J., T. Pradell, J. Pérez-Arantegui, and M. Vendrell. 2005. 'Production of lustre decoration: A comparison between ancient and modern traditional techniques', *Geoarchaeological and Bioarchaeological Studies* 3:221–24.

Morel Fatio, A., and A. Rodriguez Villa. 1876. *Henrique Cock, Relación del viaje hecho por Felipe II en 1585 a Zaragoza, Barcelona y Valencia.* Madrid: Aribau, 30–31.

Pérez-Arantegui, J., and A. Larrea. 2003. 'The secret of early nanomaterials is revealed thanks to transmission electron microscopy', *TRAC-Trends in Analytical Chemistry* 22(5):327–29.

Pérez-Arantegui, J., A. Larrea, J. Molera, T. Pradell, and M. Vendrell-Saz. 2004. 'Some aspects of the characterisation of decorations on ceramic glazes', *Applied Physics A* 79:235–39.

Pérez-Arantegui, J., J. Molera, A. Larrea, T. Pradell, M. Vendrell-Saz, I. Borgia, B. G. Brunetti, F. Cariati, P. Fermo, M. Mellini, A. Sgamelotti, and C. Viti. 2001. 'Luster pottery from the 13th to the 16th century: A nanostructured thin metallic film', *Journal of the American Ceramic Society* 84(2):442–46.

Pérez Camps, J. 1998. *La cerámica de reflejo metálico en Manises 1850–1960.* Colección Ethnos, 4. Valencia: Museo de etnología – Diputación de Valencia.

Piccolpasso, C. 1976. *I tre libri dell'arte del Vasaio, 1557.* Florence: Edizioni all'Insegna del Giglio.

Porter, Y. 2003. 'Les techniques du lustre métallique d'après le Jowhar-Nâme-Ye Nezâmi (1196 AD)', in C. Bakirtzis (ed.), *Proceedings of the Conference of Mediterranean Medieval Ceramics (Thessaloniki, 1999)*. Athens: La Caisse des Recettes Archéologiques, 427–36.

Pradell, T., J. Molera, M. Vendrell, J. Pérez-Arantegui, E. Pantos, M. Roberts, and M. DiMichiel. 2004. 'Role of cinnabar in luster production', *Journal of the American Ceramic Society* 87(6):1018–23.

Riaño, J. F. 1878. *Sobre la manera de fabricar la Loza dorada de Manises.* Madrid.

Rose-Albrecht, J. 2002. 'Aragon et Catalogne', in *Le calife, le prince et le potier : Les faïences à reflets métalliques.* Paris: Réunion des Musées Nationaux, 140–51.

CHAPTER 9

Naturam ars imitata: European Brassmaking between Craft and Science

THILO REHREN AND MARCOS MARTINÓN-TORRES

ABSTRACT *This chapter presents a summary of analytical work on medieval brassmaking crucibles, spanning more than half a millennium and tracing what is believed to be a gradual development of increasing skill and efficiency of the craftspeople who used these crucibles. This summary is then contrasted to a similarly diachronic sequence of textual sources concerned with brassmaking, illuminating the discrepancy between the matter-of-fact practitioners' reports and the somewhat befuddled attempts of philosophers of nature to understand and explain the essence of brass as opposed to copper. The different strands of knowledge generation and transfer, namely, observation and apprenticeship versus theoretical consideration and text, are placed into the changing scholarly environment from the High Middle Ages to the Renaissance. This exercise provides fresh insight not only into the development of brassmaking technology but also into the driving, or otherwise, forces in technological developments in general. The comparison of archaeological, scientific, and historical evidence is used to demonstrate the potential of such multisource studies.*

Introduction

Brass has a more chequered history than most other alloys do, so much so that despite more than three millennia of brass use, it finally was acknowledged as an alloy only less than 500 years ago. For most of its history, brass caused wonder and stimulated the minds of people trying to understand what it really was – a process that in some sense continues to this day. The reason why it took so long for brass to be recognised as an alloy has been discussed elsewhere in length (de Ruette 1995; Zacharias 1989); it suffices here to summarise the key points.

The term brass refers to an alloy of copper and zinc. Zinc concentrations in archaeological brass range from about 5 wt% at the lower end to about 30 wt% at the upper end. Practically, there is no lower limit for the zinc concentrations in copper: it is a semantic question when we start calling zinc-containing copper 'brass', and when we consider it an impure copper, or copper with just trace or low concentrations of zinc. Archaeologically, though, this is more than a linguistic question, as the term brass (rather than impure copper) strictly speaking implies an intentionally produced alloy, as opposed to an accidentally produced mixture of metals.

The problem with zinc as a metal is that at temperatures above 907°C it exists only as a vapour, but most smelting furnaces operate at much higher temperatures. Thus, metallic zinc on its own was almost unknown in Antiquity in the West (but see Rehren 1996, and Rehren in Fellmann 1999, and literature therein), ruling out the traditional way of alloying by co-fusing existing metals, as done with copper and tin to produce bronze. Instead, brass was produced by a technique called cementation, which allowed the manufacture of the copper-zinc alloy several centuries before unalloyed zinc metal could be exploited. In this process, metallic copper, charcoal, and powdered zinc ore (usually carbonates or oxides) were heated together in a crucible, so that zinc vapour formed and reacted with the copper metal, thus producing brass. Significantly, and owing to the high vapour pressure of zinc already well below its boiling point, this process can proceed at temperatures as low as 800°C. What is important here is that, contrary to the more common alloying method, only one metal is seen to enter the reaction vessel (copper), and a different type of metal is produced (brass), hence the problems in the understanding of the process.

To facilitate the reaction, the crucible had to be closed to force the zinc vapour into the copper, rather than it escaping with the fumes of the furnace. Brass with the typical Roman level of approximately 15 to 25 wt% zinc melts at 1050 to 950°C, which is well above the temperature necessary for the cementation process but still lower than the temperature required for smelting copper, of around 1100 to 1200°C. Thus, depending on the composition of the forming brass and the operating temperatures, the resulting alloy would be either solid or liquid.

The upper zinc limit for cementation brass has been intensively disputed. Empirically, very few well-dated objects are known with more than about 32 wt% zinc, although experimental work indicates that higher zinc concentrations can be achieved by both classical and modern methods of brass production (Haedecke 1973; Newbury, Notis, and

Newbury 2005; Ullwer 2001; Welter 2003; Werner 1970; Zwicker et al. 1985). This, however, has no particular implications for the perception of the material, and we are not addressing this aspect here any further.

This chapter looks at the changes in the perception of brass, as well as the understanding and explanation of the principles behind its production, taking place in Europe over the last millennium. In contrast to most papers on medieval and later brass, we do not focus on analytical data of objects themselves but juxtapose historical texts addressing the material, its properties and its production, with archaeological evidence for the latter. In doing so, we hope to identify and explain the transition from an intuitive process of improving brassmaking practice, based on tradition and practical replication with modification on the side of the practitioners, to a period of more conscious developments, based on a theoretically framed transcendence of the process on the side of scholars, and possibly involving experimental work. In addition, we attempt an insight into the evolving relationship between theory and practice in this arena, and the impact of the Renaissance rationality and empiricism in the making and understanding of brass.

Here, we are able only to scratch the very surface of this topic, which is clearly much richer archaeologically and much more complex in its textual sources than can be dealt with in the space of a few thousand words. However, we hope to show some of the potential in this subject for in-depth study.

Premedieval Brassmaking in the Old World

By way of introduction, and as a backdrop against which European brassmaking is to be understood as fundamentally different from the other alloys inherited from Antiquity, we briefly outline three main phases of brass use, mainly in Europe, separated by periods for which we have little or no evidence for brass being made or widely known.

The first phase of brass use covers the Bronze Age and Early Iron Age of the Near and Middle East. The number of brass artefacts now known from the eastern Mediterranean and Western Asia is so big that they can no longer be ignored as flukes of chance (Craddock and Eckstein 2003; Thornton and Ehlers 2003). This archaeological presence broadly correlates with textual reports in the archaic Greek literature of a particular and highly prized copper alloy called *oreichalkos*, or 'mountain copper', now widely believed to be brass (Craddock and Eckstein 2003:217). In the 1st century C.E., Pliny refers to *aurichalcum* (literally 'golden copper' but phonetically an adaptation of the Greek term) in a way suggesting that by his time this alloy was no longer being made (*Nat. Hist.* XXXIV:2–4;

translated in Rackham 1952). In the absence of any published archaeological evidence for the production of this alloy in pre-Roman times, and given its typically low and rather variable zinc content, one may assume that it was produced by smelting a particular zinc-rich copper ore. This assumption is consistent with the use of a specific term for this metal, identifying it as a variant of copper but sufficiently different for it to not be included in the generic *chalkos* (*aes* in Latin) used for copper and bronze. It is also consistent with the archaeological disappearance of this alloy, possibly because of simple exhaustion of the particular ore deposit(s) from which it was smelted. At least during the final phase of this period, in the Iron Age, *oreichalkos* seems to have enjoyed a reputation as a rare and desirable material, worthy of use by the best of the time, mostly for jewellery and ornaments.

The second phase in this history spans from the mid-1st century B.C.E. up to the late Roman period. It starts with the sudden appearance of small quantities of brass both in the East and to the north and west of the Roman Empire, followed by a boom of brass use, first in military implements and other state-controlled contexts such as currency, during the late first century B.C.E. and the first century C.E. (Bayley 1998; Ponting 2002; Ponting and Segal 1998; Weeks 2004). This gradually develops into a period of more widely distributed, and literally also diluted, use, when brass and zinc-containing ternary alloys permeate into general domestic use. The supply of fresh brass during this later stage of the second phase seems to have dried up, as no or only very little high-zinc, high-quality brass is still being used then (Bayley and Butcher 2004; Dungworth 1996, 1997a, 1997b).

There is a considerable body of literature on alloy compositions from this period (see references above), but only very few instances of brass-making installations have been published so far. These have in common that they clearly point toward the conscious production of brass in closed crucibles using the cementation method, enabling relatively close control over the resulting alloy composition. Those few case studies reported so far paint a rather diverse picture, ranging from tiny purpose-made acorn-shaped (and -sized!) crucibles used in Colonial Ulpia Traiana/Xanten (Rehren 1999a) to large amphorae-like vessels used for brass-making in Lugdunum/Lyon (Picon, Le Nezet-Celestin, and Desbat 1995) and normal, fist-sized, metallurgical crucibles in Roman Britain (Bayley 1998) (Figure 9.1). These vessels are all conspicuously free of slag, suggesting that the process was probably done at a relatively low temperature, possibly involving only solid copper (but see Craddock and Eckstein 2003). The resulting brass, thus formed in the solid state, would probably be melted and further refined only when needed for casting. The cementation crucibles were lidded and tightly sealed, and the pressure released

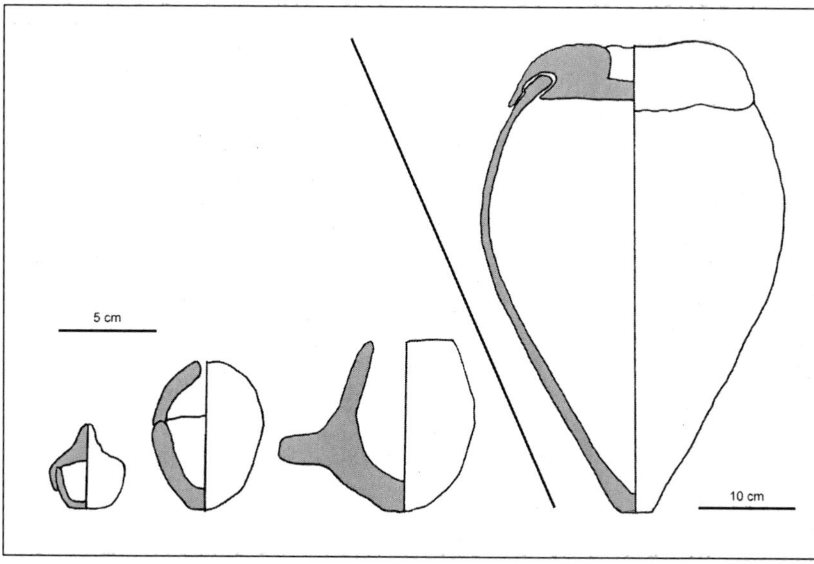

Figure 9.1 • Formal diversity in Roman brassmaking crucibles; from left to right, examples of crucibles from Xanten (Germany); Culver Street in Colchester (United Kingdom) (© Justine Bayley); Palace Street in Canterbury (United Kingdom) (© Justine Bayley); and Lyon (France) (© Société Française d'Étude de la Céramique Antique en Gaule). Note the different scale of the Lyon crucible, which would otherwise appear even larger when compared to the rest; see text for references (all crucibles redrawn by M. Martinón-Torres).

probably through the porous ceramic fabric and, in some cases, though a hole in the lids. Remarkable is the wide range in vessel sizes between the three more or less contemporary published production sites mentioned above, indicating that a degree of variability existed between different production centres, while the basic principle was the same for all three.

The systematic and consistent production of brass of a certain quality clearly demonstrates full mastery of the process during the early part of this second phase of brass use, while textual references show that brass was understood as a derivative of copper. Unfortunately, Pliny does not discuss its production, although he details that 'Livian' copper from Gallia/Gaul is better suited to be treated with *cadmea* (zinc ore) than that from other sources and produces a metal almost as good as the traditional *oreichalkos* (*Nat. Hist.* XXXIV:2–4, 100–103; translated in Rackham 1952). Discussion of the reason(s) for the disappearance of freshly made brass during the later Roman Empire is beyond the scope of this paper (cf Dungworth 1996). Pliny's failure to report any technical details may indicate that the practicals of brassmaking were not in the public domain and that it was a loss of know-how or state-controlled demand rather than exhaustion of raw materials that led to its demise. Hence, we have a fair idea of the production and perception of brass as a relatively cheap copper-related metal during the Roman period but no feeling whether the Roman or indigenous brassmakers considered their craft different from that of other alloy makers, nor why its production in the West seemingly ended well before the collapse of the Roman Empire.

Medieval and Early Modern Brassmaking in Europe

The third phase of brass use in Europe begins in the High Middle Ages and continues to this day. For this phase, we have the most comprehensive archaeological and textual evidence, and the emphasis of this chapter is very much on the first half of this third phase, that is, roughly from 1000 to 1600 C.E. Brass was again being made by cementation, soon replacing bronze as the alloy of choice. It is this classical method of brass production, prevailing in Europe and the Middle East until the Industrial Revolution, that led to so much confusion about the nature of brass.

As noted above, metallic zinc occurs in this process only ever as a vapour phase within the reaction vessel, invisible to the metal worker of the time and unknown to the scholars interested in the materials of nature. Accordingly, the actual amount of zinc entering the copper was less well-controlled than in other alloys. Furthermore, this peculiarity raised serious doubts whether brass really qualified as an alloy or whether it was just a particular type of copper 'coloured' yellow. It is this enigmatic change in the properties of copper, similar to the change of properties known from 'proper' alloying when making bronze, but not linked to any known metal, that puzzled scholars for many hundreds of years. However, this uncertainty did not preclude generations of metal workers from mastering and perfecting the process, so much so that brass quickly became the dominant copper alloy of the medieval and the modern periods, extensively used for everyday life implements as much as for secular and ecclesiastical works of art – as widely documented in analytical studies (for example, Cameron 1974; Caple 1995; Hachenberg 2004, 2006; Mitchiner, Mortimer, and Pollard 1986; Pollard and Heron 1996; de Ruette 1996; Riederer 1988).

The six centuries of brass use starting around 1000 C.E. are of particular interest here, since they provide both a unique sequence of archaeological evidence for brassmaking and a series of highly illuminating texts about the nature of this particular alloy, written by some of the brightest minds of their time. In the following text, we summarise the archaeological evidence for the gradual evolution of brassmaking in Central Europe, from the late 1st millennium C.E. up to the Renaissance. This summary will then be contrasted with some of the contemporary sources discussing the nature of this alloy and its production, with the aim to understand the changing perception of brass. Eventually, we look at the relationship between the practical mastery and the theoretical understanding of brassmaking, in an attempt to judge the degree of mutual dependency of practitioners and theoreticians in the field.

The Evolution of Brassmaking: Archaeological Evidence

As noted above, the third phase in the history of European brassmaking starts after the mid-1st millennium C.E., with fresh brass emerging in Scandinavian and Viking contexts (Eremin, Graham-Campbell, and Wilthew 2002). By the time of the Carolingian revival, brass is the main copper alloy in use, and it will retain this position to this day. Thus, brass was and is clearly an alloy of huge economic significance, and its production and use played a central role in many aspects of the development of the modern world. Little, however, do we know of the early period of its production, and the source of the Viking-period brass remains enigmatic.

The first archaeological evidence for brassmaking, from Dortmund in western Germany, dates to the late 1st millennium C.E., followed by a series of somewhat later instances excavated across Germany. The evidence from Soest and Schwerte, some 50 km to the east of Dortmund, dates to the first half of the 2nd millennium C.E. The latest evidence of relevance here is from Zwickau in southeast Germany, dating to the late 15th century C.E.

Of interest to us is the evolution of the crucible design over the period of half a millennium, particularly in size and profile. On the contrary, we do not address in any details the zinc ores and their preparation, for which the existing information is even scantier and has been summarised elsewhere (Craddock and Eckstein 2003). It suffices to remember that two main types of zinc ores were used: on the one hand, 'natural' minerals in the form of calcined zinc carbonate ($ZnCO_3$) or roasted sphalerite (ZnS) – generally called calamine, *cadmia fossilis*, *galmey*, and so on, depending on the authors; on the other hand, 'artificial' zinc oxide-rich crusts scraped as byproducts from the inner walls of furnaces where zinc-bearing lead ores were smelted – named furnace calamine, *cadmia fornacum*, tutty, and others (cf de Ruette 1995).

The Dortmund crucibles are short and cylindrical in shape, about 6 to 8 cm in diameter, and usually less than 10 cm high, corresponding to a volume of less than half a litre (Figure 9.2). Little if any evidence for a lid exists for these vessels, suggesting a rather wasteful process with a high proportion of the zinc vapour escaping from the reaction vessel instead

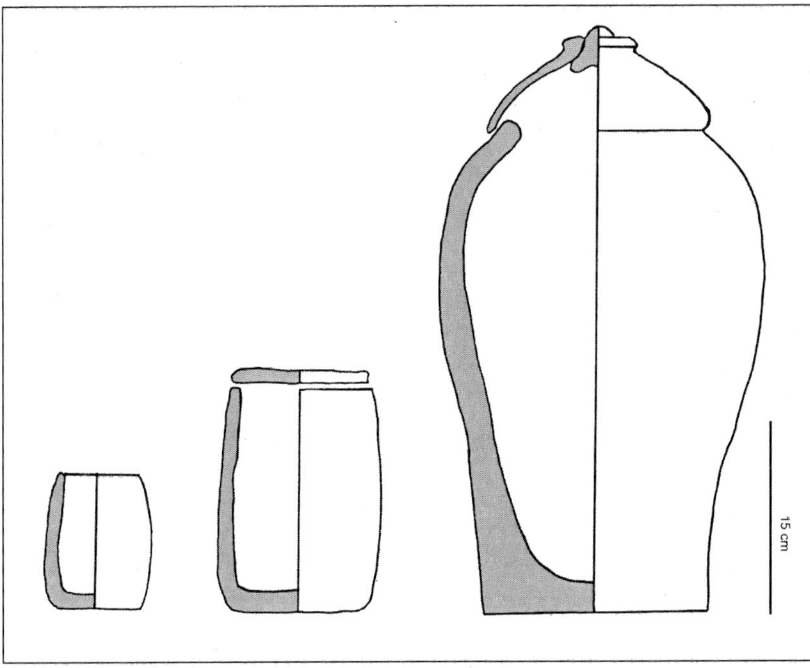

Figure 9.2 • Medieval and Renaissance brassmaking crucibles from Germany; from left to right, in chronological sequence, sketch drawings of crucibles from Dortmund; Soest/Schwerte; and Zwickau; see text for references.

of being absorbed by the copper to form brass (Rehren et al. 1993). The walls are around 1 cm thick, their fabric is dense, and, with over 27 wt% Al_2O_3 (normalised data after Martinón-Torres 2001; Rehren 1999b), much more alumina-rich and hence more refractory than the contemporary local domestic wares.

The Soest and Schwerte vessels, roughly thought to be two to three centuries more recent, are considerably bigger than the Dortmund crucibles. They are tubular with a round bottom, an internal diameter of around 9 cm and a height of at least 15 to 20 cm (Figure 9.2). The ceramic is visually almost indistinguishable from the earlier Dortmund crucibles and, although less rich in alumina (19 wt% Al_2O_3; normalised data after Rehren 1999b) still sufficiently refractory to serve its purpose. Significantly, though, there are several ceramic disks among the assemblage from Soest which were tentatively identified as lids. These would have to some extent limited the zinc loss, even though they probably only loosely fitted the mouths of the crucibles.

In all these cases, the evidence is for a cementation process as in the Roman examples, but invariably involving slag formation within the crucibles. This suggests that the operating temperatures were now on the whole higher than in the Roman period and probably involved melting the brass in the crucibles (Rehren 1999b).

Another several hundred years later, in the late 15th century C.E., brassmaking vessels had apparently moved up further in scale and technical design. At Zwickau, brassmaking vessels were by now holding about 20 kg of metal, having a height of at least 40 cm and a body with a bulbous profile reaching 28 cm in diameter, resulting in a volume of approximately 12.5 litres (Martinón-Torres and Rehren 2002) (Figure 9.2). They are made from a ceramic of only medium alumina content (around 17 wt%), making them less refractory than the earlier vessels from Dortmund and similar to those from Soest and Schwerte. However, the mechanical stability of these large vessels was ensured by walls that are rather thick, at around 2.5 cm, and further covered with an outer layer of even less refractory ceramic – an established method of improving performance of technical ceramics (Rehren 2003). Furthermore, the openings of the Zwickau vessels were covered with a sophisticated dome-shaped lid, itself comprising a smaller opening that was closed with a lump of clay if and when needed. While a thin film of slag appears inside some of these vessels, this is a reaction layer between the ceramic fabric, zinc oxide, and an iron oxide-rich gangue, but it appears that the brass did not melt during the cementation process, and there is evidence, within the same assemblage, for the melting and refining of fresh brass in different, triangular crucibles (Martinón-Torres and Rehren 2002).

Thus, overall we see a growth in crucible volume from the small cups of Dortmund to the buckets of Zwickau, with the ceramic design being adapted to match the physical demands of the increased weights of the charge (Figure 9.2). In addition, the opening of the crucibles and their lids become much more sophisticated, obviously in an attempt to minimise loss of zinc vapour while still maintaining access to manipulate the charge with iron rods (see below). Finally, the later crucibles indicate a better control of the working temperatures, which would result in a more economic use of the raw materials and higher-quality brasses. Altogether, it is tempting to picture a technology being passed on from one generation of craftspeople to the next, whereby brassmaking improves by experience and through iteration but no significant innovation or breakthrough is noticeable.

Scholarly Understanding of Brassmaking? Historical Evidence

The Middle Ages

The earliest detailed description of brassmaking is by Theophilus, dating to the 12th century C.E. and therefore roughly contemporary to the evidence from Soest und Schwerte. In his third book of *On divers arts*, he advises:

> And when the crucibles are red-hot take some calamine, about which I spoke above, that has been [calcined and] ground up very fine with charcoal, and put it into each of the crucibles until they are about one-sixth full, then fill them up completely with the above-mentioned copper, and cover them with charcoal [. . .] Now, when the copper is completely melted, take a slender, long, bent iron rod with a wooden handle and stir carefully so that the calamine is alloyed with the copper. Then with long tongs raise each crucible slightly and move them a little from their position so that they may not stick to the hearth. Put calamine in them all again as before and fill them with copper and cover them with charcoal. When it is once more completely melted, stir again very carefully and remove one crucible with the tongs and pour everything into little furrows cut in the ground. Then put the crucible back in its place. (Hawthorne and Smith 1979:143–44)

The chronologically next source, discussing the nature of brass as well as its production, is the *Book of Minerals* by Albertus Magnus, written probably in the early second half of the 13th century. In Chapter 6 of the fourth book he explains the nature of copper as a result of the combination of the particular qualities and quantities of quicksilver and sulphur that make up all metals. For him,

> Quicksilver is good, not full of dross and dirt, but still not completely cleansed of extraneous moisture; and [. . .] the substance of the Sulphur is full of dross, burning hot and partly burnt, and in this condition it is mixed with the Quicksilver, both in substance and in quality. Then undoubtedly it changes the copper to a red colour; and because neither [the Sulphur nor the Quicksilver] is sufficiently subtle, they cannot be well mixed. And this will make copper, which is not at all well mixed, since much dross is separated from it, and it evaporates greatly in the fire. (Wyckoff 1967:223)

He then goes on to explain how and why the copper is not well mixed and burns, concluding that: 'Now, therefore, we understand the material of copper; it is a metal having rather more Quicksilver than it ought to have, which has been converted into a red form by mixture with burning Sulphur' (Wyckoff 1967:224). Building on this understanding of copper, he then explains the making of brass as follows:

> But those who carry on much work with copper in our region – that is, in Paris and Cologne and other places where I have been and seen this tested by experience – convert copper into brass [*aurichalcum*] by means of a powder of a stone called calamine. And when this stone evaporates there still remains a dark lustre, approaching the appearance of gold. And to make it paler in colour, and so more like the yellow of gold, they mix in a little tin;

> but because of this, brass loses the malleability of copper. And those who wish to deceive and to produce a lustre like gold 'bind' the stone so that it may remain longer in the copper on the fire, and not evaporate from it so quickly. And the 'binding' [is done] with 'oil of glass'. They take fragments of glass, crushed and sprinkled into the crucible on the copper after the calamina is put in; and then the glass that has been put in floats on the top of the copper and does not allow the power of the stone to evaporate, but reflects the vapour of the stone down into the copper. And in this way the copper is thoroughly purified for a long time and the drossy material in it is burnt up. But after a while the oil of glass evaporates, and then the power of the stone evaporates, too; but the brass is made much more brilliant than it would have been without it [. . .] But Hermes says that if powdered tutty is mixed with molten copper – either white tutty or red – it changes the copper to the colour of gold. What tutty is will be explained in the following book, where 'intermediates' are treated. But it is enough [to say] here that the burning heat of tutty consumes the earthiness and purges the superfluous moisture out of the copper; and so then it will be more beautiful. But the power of tutty, too, evaporates if it stands for a long time on the fire; and therefore, unless some remedy is used, the tutty will evaporate and the copper will regain its original colour. (Wyckoff 1967:224–25)

These two texts, separated by only about a century, show two radically different approaches to the topic. Theophilus gives a very matter-of-fact description, a recipe, of how to make brass; he does not seem to spend much thought on what actually happens to the copper in the process. Clearly, brass is an important material to him, and he spends several paragraphs on the best preparation of the furnace for brassmaking, as well as on the different qualities of brass: qualities that are of relevance if the metal subsequently is to be gilded or hammered rather than cast. Based on this text, any dexterous person can successfully make brass, regardless of whether they have a degree in science, or philosophy, or are barely literate enough to read his text. No word is spared on theoretical speculations. Thus it is not surprising that the contemporary archaeological evidence from brass workshops is much more easily explained in the light of this text.

Albertus Magnus, in stark contrast, explains in much more detail the basic constituents that make up the various metals, manifested in their different degrees and qualities of coolness, humidity, and 'burning heat', and as embodied in the principles for the generation of all metals: quicksilver (mercury) and sulphur. He then places copper within this theoretical framework, which he carefully developed from the scriptures of the classical authors, before addressing brass. The relevant sentences here are an interesting mixture of plain observations of current practice in Paris and Cologne and attempts to accommodate what he noticed within his theoretical framework of varying 'qualities' of metal that have to be 'purged' of inferior aspects.

It would be quite difficult to follow his description in practice, though. For instance, no mention is made by him of the crucial ingredient charcoal, necessary for the zinc ore to be reduced to metal vapour and without which the whole cementation process would not work. However, Albertus devotes many words to discuss the addition of crushed glass as a protective coating for the melt, which, from a present-day perspective, is a less important aspect of the process. The ensuing discussion demonstrates some form of practical understanding that the central component zinc is volatile and should be kept from evaporating. However, the agent that he describes to evaporate, and that he clearly links to the change in colour from red copper to yellow brass (and back to red copper through evaporation), is seen more as an immaterial quality, a 'power' of *tutty*, rather than substantially the *tutty* itself – which is, of course, zinc oxide.

The apparent discrepancy could not be stronger, but it can be somewhat explained by considering the backgrounds and aims of both authors. Theophilus was a Benedictine artisan monk extensively experienced in arts and crafts, he collected practical recipes as a memory aid, and showed no concern with explanations. The result was a book that, from a technical viewpoint, was not superseded 'until the books by Cennini (1437) on painting, Månsson (ca 1520) on glass, and Biringuccio (1540) on almost everything but painting' (Hawthorne and Smith 1979:xxxi).

Albertus Magnus, widely acknowledged as one of the greatest thinkers and most prolific writers of the Middle Ages, held high office in the Dominican Order and later served as a bishop, and was eventually canonised in 1931. From a higher social rank and in a completely different academic pursuit, what he attempted was an explanation for the nature and formation of the natural world that could reconcile Aristotle's four elements, Avicenna's sulphur-quicksilver theory, further alchemical beliefs, and even Christian theology. Returning to brassmaking, his main objective was not to produce a beautiful metal of appropriate material properties but to investigate the causes behind its formation and, ideally, to extract relevant information that would aid the search of metallic transmutation.

Medieval alchemists sought the 'tincture' or 'elixir' that, added to base metals, would give them the quality – that is, the actuality – of gold. It was in this manner that they sought to imitate natural gold with artifice. The sensorial qualities of artificial brass did indeed compare reasonably well those of natural gold, but, to the alchemists' despair, the yellow colour of brass (that is, the zinc) evaporated on prolonged melting. In another text attributed to Albertus, several paragraphs insist that 'the alchemical art requires that some means be discovered whereby the spirit does not escape, is not consumed or burned, but instead penetrates and is mixed with all parts of the metal, so that it may tint with a permanent tincture' (Kibre 1944:305). In these pages, *tutia* (zinc ore), *marcasita* (iron sulphide), and *sale armoniaco* (ammonium chloride) – all of which contain volatile elements – are mentioned as potential 'tinctures' that need to be processed and improved.

It is probably within this context that Albertus calls 'deceivers' (see above) those who sprinkle broken glass in the crucibles to prevent evaporation, since this does not solve the problem of the 'permanent tincture'. Avoiding colour losses was good enough for a metallurgist, but not acceptable for a demanding scholar.

All in all, these written sources demarcate two distinct traditions: on the one hand, those such as Theophilus, concerned with the practicalities; on the other hand, those pursuing explanations, as was the case of Albertus Magnus. Not only Theophilus but also Albertus had some first-hand acquaintance with metallurgical practice – probably obtained in the Rammelsberg some 30 years before writing his book (Wyckoff 1967). His theoretical speculations bear some technical basis, but craftsmen and scholars are working in different spheres. Meaningfully, Albertus disregards Pliny because 'he does not offer an intelligent explanation of the causes common to all stones' (Wyckoff 1967:10), and nothing in *The Book of Minerals* suggests that Albertus, the great encyclopedist, ever read Theophilus.

The Renaissance

The 16th century saw a great expansion in printing technology, together with other technological and scientific achievements, and a fervent humanistic interest in understanding nature and recording knowledge. In this context, several major treatises on mineralogy and metallurgy appeared, where we find detailed accounts on brassmaking. Three crucial texts must be considered here: Vannoccio Biringuccio's *De la pirotechnia*, first published in 1540, Georg Agricola's *De natura fossilium* – notably the second edition of 1558 – and Lazarus Ercker's 1580 *Beschreibung aller fürnemisten Mineralischen Ertzt vnnd Berckwercksarten*. How do they differ from the earlier texts?

Biringuccio gives a brief description very similar to Theophilus's, although he fails to mention the necessary charcoal charge:

> For this process they placed in each one of the vessels 25 pounds of German rosette copper broken in pieces, and they filled up the rest to within 2 dita of the rim with a powder of a mineral earth, yellowish in color and very heavy, that they called calamine. The rest of the empty space in the crucible they filled with powdered glass [. . .] Then they applied a melting fire for 24 hours and after this time they found the material entirely fused, and the

copper, which was red before, had become a smooth and lovely yellow, almost like 24-carat gold in color. (Smith and Gnudi 1990:72)

His approach to, and understanding of, the matter is clearly expressed in the introductory paragraph (Smith and Gnudi 1990:70):

Since I have no knowledge other than that gained through my own eyes, I can tell you as a certainty that just as steel is iron converted by art into almost another kind of metal, so also brass is copper given a yellow color by art.

Notably, he describes the furnaces used, and he also mentions that 'little fitted clay shutters' may be used on the crucibles (Smith and Gnudi 1990:72). Later he addresses the nature of this 'earth that colors copper into brass':

I do not know that this earth serves any other purpose than coloring copper, because the mineral matter is of bad elemental mixture and poorly fixed [. . .] and not only does it color copper another color, but it increases its volume so much that the workman covers the cost of the copper and the expense of coloring it. (Smith and Gnudi 1990:75)

Further, he elucidates:

I believe that in its nature it is of a hot and dry quality like marcasite, as experience shows, because it does not melt alone by itself but burns and all its substance goes off in smoke. (Smith and Gnudi 1990:113)

Soon after the publication of *De la pirotechnia*, in 1546, the first edition of Agricola's *De natura fossilium* came out. Unlike his famous compendium *De re metallica* (Hoover and Hoover 1950), the former volume addressed brass production. This discussion was further expanded in the second edition, published in 1558 (text on brass translated and discussed in Martinón-Torres and Rehren 2002). Agricola starts his exposition by referring to Pliny and the old *oreichalkos*, before restating an old alchemical principle: *naturam ars imitata* ('art imitates nature') – it is thus possible to artificially dye copper. He then presents two alternative ways of making brass. The first one is again similar to Biringuccio's, although he notes that the charge is best arranged in the crucible in layers, recaptures the possibility of adding crushed glass, and explicitly acknowledges that either mineral or furnace calamine may be used. He also insists on the need to carefully adjust the working temperatures. In the second method, the cementation principle remains the same, but he provides specific details such as the use of large pots, covered with a perforated lid and externally coated with another clay layer – almost exactly the description of the archaeological crucibles from Zwickau. It is also worth noting that he quantifies the amounts of charge to be used and, particularly, the fact that the copper, 'transformed into brass [. . .] has become much heavier' (Martinón-Torres and Rehren 2002:102).

A few decades later, Lazarus Ercker provides another very detailed account of brassmaking. He reports that the calamine is first roasted, then mixed with twice its volume of charcoal dust and dampened, with water, urine, or alum solution. After a blow-by-blow recipe of how to charge and fire the crucibles, he notes that the amount of copper charged in the crucibles increases in weight over the course of nine hours of firing, from 64 units to 90 units. Again, he judges the quality of the product by its 'fine colour', but all of his discussion centres on mundane issues such as the comparison between natural calamine from the ore deposits and artificial calamine collected from the furnaces near Goslar, and the lead content of the various types (Sisco and Smith 1951:255–56). His record is accurate and comprehensive, from the practice for the practice and seemingly with little regard for theoretical speculation and systematic ordering.

* * *

All three accounts have much in common: for example, the use of colour as a measure for the quality of the metal produced, but also the recognition of the increase in mass – be it economically quantified by Biringuccio ('covers the cost of the copper and the expense of coloring it') or directly measured as weight gain by Agricola and Ercker. Some differences do occur, though these are more subtle than those identified above between Theophilus and Albertus. Agricola, on the one hand, was a foremost scholar of his time, with studies of philology, philosophy and science, and a degree in medicine. As a humanist, he wrote in Latin for a learned audience and filled his text with references to the classical authors. Biringuccio and Ercker, on the other hand, were primarily trained as practitioners – the former as a founder, the latter as an assayer – and wrote in their vernacular languages. However, they were also exposed to broader institutional and intellectual arenas: both of them spent their lives under the patronage of wealthy families and worked alongside scholars and scientists.

Perhaps most remarkable is the fact that, by and large, these authors are providing clear and effective metallurgical recipes, but also some observations and thoughts on the nature of the elements involved and the process taking place. The emphasis certainly is on the practical side, but, as exemplified in Biringuccio's lines on the calamine – and many others not reported here – they show an awareness of existing theoretical speculations, which they combine with their own observations – such as the increase in weight of brass. They are writing in the wake of the Renaissance: the world is being explored and conquered, information is published and discussed, knowledge about the natural world is keenly sought, and, increasingly, practical observations and theoretical abstractions are being linked. The boundaries between the scientist and the craftsman, between the alchemist and the metallurgist, between theory and practice, are more and more blurred (Beretta 1997; Long 1991, 2001; Martinón-Torres and Rehren 2005). This is epitomised by the *'naturam ars imitata'*, which inspires Agricola as much as it inspired Albertus and others. Of course, some are still concerned with perfecting the 'permanent tincture' for 'brass, which is the Philosopher's Gold, and that is true' (att. Alfonso V King of Portugal 1651:69), but they all share a more direct connection among experiment, abstraction, and practice. To what extent did this Renaissance empiricism and rationality affect the evolution of brassmaking?

Interestingly, Ercker not only describes brass production in close proximity to the smelting area of Goslar. In his *Brief Report on the Rammelsberg*, first published in 1565 (Beierlein 1968), he describes the occurrence of a strange white-metallic material that seeps out of some of the lead-smelting furnaces. This material, which he says is called *Contrafeth*, is most likely metallic zinc, condensing in cracks and crevices in the furnace walls. He suggests that much more of it could be made if one only spent some thought on it and laments that the workmen are not interested in new things, even if they would be useful ('Such *contrafeth* could be made in quantity if effort and thought were put to it; but it is not valued, and the worker and smelter put no effort into new things, however much it would help'; translated from Beierlein 1968:252). He goes on to say that it is not regularly recovered and only ever collected when specifically requested by someone paying them a tip. Furthermore, he states that nothing can be made of this metal on its own, because it is too brittle, and proposes to alloy it with other metals to make it useable. Here we see, for the first time, the expression of the belief that a new material could be developed – through experiment and thought – into an economically interesting commodity: the beginning of industry-based research and development in the modern sense. It is also during this period that ingots of metallic zinc begin to arrive from China and India where, contrary to Europe, the distillation of zinc had been mastered centuries before. This puzzles scholars such as Andreas Libavius, who describe it as a 'kind of tin' (de Ruette 1995:195).

Eventually, experimentation and active research led to 'speltering', the process in which metallic zinc is alloyed with copper to produce brass. This possibility seems to have been first realised in Europe by the 17th-century German chymist Johann Glauber, who wrote in 1657:

> Zink is a volatile mineral, or a half ripe metal when it is drawn out of its ore. It is much clearer and brighter than tin, yet not so malleable and fluxile as tin is [. . .] We have it not much growing in Germany, but great quantity of it is every year brought us by the merchants out of the East Indies [. . .] It is a golden but an unripe mineral, it gives red copper a yellow colour and turns it into brass, as lapis calaminaris doth; and indeed that same stone is nothing else but unmeltable zink, and this zink may properly be called a susile lapis calamnaris (*sic*); for as much as both of them partake of one nature. (translated in Packe 1689:319)

The realisation was there. However, it would be another 200 years before the true nature of zinc as a metal was fully understood and its production sufficiently perfected to replace cementation as the basis of industrial brass production.

Discussion

It is tempting to see in the few examples of archaeological evidence for brassmaking briefly presented above a chronological trend line from small and primitive, unlidded vessels to successively larger and better closed vessels, representing an evolution of increasing productivity and improved procedures. Minor but significant design details, such as the wall thickness and the application of an outer wrap of less refractory clay, together with the better temperature adjustment in Zwickau, also point in this direction. They could well have been the result of semicontinuous minute changes in the practice, leading over 500 years to significant advances. This, however, is only the positivistic interpretation of the scenario. As seen from the three Roman examples mentioned above, it is as well possible that different crucible sizes coexisted (almost) simultaneously, possibly reflecting different economic settings, scales of production, or simply local technological styles rather than necessarily representing an evolutionary sequence. Only a much larger and better contextualised archaeological sample will shed clearer light on this issue.

The textual evidence draws a more complex picture, as so often in this period. At a first glance, one sees a fault line running through the corpus of medieval texts addressing brass and brassmaking. On the one side stand scholars such as Albertus Magnus, who primarily try to understand fundamentally the material world using theoretical concepts inherited from the classical authors and further developed by their own thoughts. On the other side are writers more concerned with description rather than explanation, who carefully report their observations and convey recipes to facilitate the arts and crafts, rather than philosophy.

In the Renaissance, the texts suggest a more systematic combination of practice and abstract thought that enabled a more practical understanding of the matter. This required, first, realising that the cementation process entailed a significant weight increase and not just a colour change. This observation is a major step forward from the old concept of a 'power' that colours copper an alien colour, shifting the balance from the ideas of hotness, inherent moisture, and so on to acknowledging that a physical substance is being added to the copper and effects this change. A second important realisation will be the fact that this extra substance is indeed zinc and that brass can therefore be made by simply alloying the two metals.

Much needs to be said about the transmission of knowledge, both over time and between the two sides – practitioners and scholars. It would appear that the latter owe the former a lot of careful observation and experimentation; however, practitioners such as Biringuccio and Agricola explicitly credit 'the alchemists' with the invention of brassmaking. Whether this is a reference to a possible 'Eastern' origin of the revived medieval brassmaking or simply a generic attribution for everything enigmatic and artful is another question outside the remit of this chapter.

Whatever the case, the historical record shows that, in theory, by the 17th century everything was ready for a Renaissance breakthrough in brassmaking that could have turned it into an even more efficient and profitable industry – hence Ercker's encouragement of research and development. What happened in practice? Did these seeds of innovation grow?

To cut the story short, it is enough to remember that a method for the industrial production of zinc by distillation was first patented only in the late 1730s (by William Champion in Bristol, England), and still this proved rudimentary and unprofitable, and in need of much development, before it could allow speltering as an established brassmaking method (Day 1973, 1991, 1995, 1998; Dungworth and White 2007). Not only for Britain but also for Bavaria (Priesner 2000) we have a rich historical and archaeological record evidencing that the industrial production of brass continued to rely on the traditional cementation method well into the 19th century, that is, over 200 years after the necessary discoveries for direct alloying had taken place.

There is not enough space to elaborate on the reasons why this Renaissance wave of experimentation and discovery took so long to affect daily, industrial practice, that is, what we would now term the relationships between science and technology. In considering possible explanations, we are inclined to agree with Johann Glauber: in spite of his excitement about, and encouragement for, the exploitation of the new metal zinc, 'which is most excellently excellent (*sic*)', he was frustrated that 'men are hardly drawn back from an old custom' (Packe 1689:320). At a practical level, tradition was clearly stronger than the stimuli for innovation, hence the slow pace of industrial applied research. Surely, additional reasons for this lie in the particular socioeconomic contexts, the balance between the perceived costs and benefits, competition between companies, and the flow of information between scholars and practitioners. As this chapter has tried to show, the full picture will be reconstructed only through combined work in archives and libraries, archaeological sites, and archaeometric laboratories.

Acknowledgments

This chapter grew out of many years of cooperation with many colleagues, allowing access to excavated finds as well as to their thoughts. We can mention only some of them, including N. Zieling, H. Brink-Kloke, and J. Beutmann for their archaeological expertise, and Chr. Bartels and M. Charlton for their thoughts and comments on ideas developed in this text. Any errors of judgment remain ours. The underlying analytical work, published elsewhere, was done first in the Institut für Archäometallurgie at the Deutsches Bergbau-Museum, Bochum, and later at the Wolfson Archaeological Science Laboratories at the UCL Institute of Archaeology, London. We are indebted to these institutions and their staff, particularly Andreas Ludwig and Kevin Reeves, for their unfailing support. We acknowledge with gratitude financial support given to MMT by the Barrié de la Maza Foundation (Spain) and the British Council (UK).

References

Alfonso V, King of Portugal. 1651. *Five Treatises of the Philosopher's Stone: Two of Alphonso King of Portugall, as It Was Written with His Own Hand, and Taken out of His Closset*... London: Thomas Harper.

Bayley, J. 1998. 'The production of brass in Antiquity with particular reference to Roman Britain', in P. T. Craddock (ed.), *2000 Years of Zinc and Brass*. British Museum Occasional Paper 50. London: The British Museum, 7–26.

Bayley, J., and S. Butcher. 2004. *Roman Brooches in Britain: A Technological and Typological Study Based on the Richborough Collection*. London: The Society of Antiquaries of London.

Beierlein, P. R. 1968. *Lazarus Ercker: Drei Schriften*. Bochum: VFKK.

Beretta, M. 1997. 'Humanism and chemistry: The spread of Georgius Agricola's metallurgical writings' *Nuncius: Annali di Storia della Scienza* XII:17–47.

Cameron, H. K. 1974. 'Technical aspects of medieval monumental brasses', *Archaeological Journal* 131:215–36.

Caple, C. 1995. 'Factors in the production of medieval and post-medieval brass pins', in D. R. Hook and D. R. M. Gaimster (eds.), *Trade and Discovery: The Scientific Study of Artefacts from Post-Medieval Europe and Beyond*. British Museum Occasional Paper 109. London: The British Museum, 221–34.

Craddock, P. T., and K. Eckstein. 2003. 'Production of brass in Antiquity by direct reduction', in P. T. Craddock and J. Lang (eds.), *Mining and Metal Production Through the Ages*. London: The British Museum Press, 216–30.

Day, J. 1973. *Bristol Brass: A History of the Industry*. Newton Abbot: David and Charles.

Day, J. 1991. 'Copper, zinc and brass production', in J. Day and R. F. Tylecote (eds.), *The Industrial Revolution in Metals.* London: The Institute of Metals, 131–99.

———. 1995. 'Trade and innovation in non-ferrous metals at Bristol', in D. R. Hook and D. R. M. Gaimster (eds.), *Trade and Discovery: The Scientific Study of Artefacts from Post-Medieval Europe and Beyond.* British Museum Occasional Paper 109. London: The British Museum, 205–20.

———. 1998. 'Brass and zinc in Europe from the Middle Ages until the nineteenth century', in P. T. Craddock (ed.), *2000 Years of Zinc and Brass.* British Museum Occasional Paper 50. London: The British Museum, 133–58.

de Ruette, M. 1995. 'From *conterfei* and *speauter* to zinc: The development of the understanding of the nature of zinc and brass in post-medieval Europe', in D. R. Hook and D. R. M. Gaimster (eds.), *Trade and Discovery: The Scientific Study of Artefacts from Post-Medieval Europe and Beyond.* British Museum Occasional Paper 109. London: The British Museum, 195–203.

———. 1996. 'Brass foundry workshops of the Southern Low Countries and the Principality of Liége: A technical approach', *Historical Metallurgy* 30:116–20.

Dungworth, D. 1996. 'Caley's "zinc decline" reconsidered', *Numismatic Chronicle* 156:228–34.

———. 1997a. 'Iron Age and Roman copper alloys from northern Britain', *Internet Archaeology* 2, http://intarch.ac.uk/journal/issue2/dungworth/index.html.

———. 1997b. 'Roman copper alloys: Analysis of artefacts from northern Britain', *Journal of Archaeological Science* 24:901–10.

Dungworth, D., and H. White. 2007. 'Scientific examination of zinc-distillation remains from Warmley, Bristol', *Historical Metallurgy* 41:77–83.

Eremin, K., J. Graham-Campbell, and P. Wilthew. 2002. 'Analysis of copper-alloy artefacts from pagan Norse graves', in K. T. Biró and E. Jerem (eds.), *Archaeometry 98: Proceedings of the 31st Symposium, Budapest, April 26–May 3 1998.* BAR International Series, 1042. Oxford: Archaeopress, 343–49.

Fellmann, R. 1999. 'Das Zinktäfelchen vom Thormebodewald auf der Engehalbinsel bei Bern und seine keltische Inschrift', *Archaeologie im Kanton Bern* 4B:133–75.

Hachenberg, K. F. 2004. 'Der Werkstoff Messing im Musikinstrumentenbau vom 16. bis zum Ende des 18. Jahrhunderts', in B. E. H. Schmuhl and M. Lustig (eds.), *Jagd- und Waldhörner: Geschichte und musikalische Nutzung. 25. Musikinstrumentenbau-Symposium, Michaelstein, 8. bis 10. Oktober 2004,* Michaelsteiner Konferenzberichte 70. Augsburg: Wißner.

———. 2006. 'Die analytische Beschaffenheit des Nürnberger Messinggusses von 1450 bis um 1720', *Guss im Wandel der Zeit* 39:11–9.

Haedecke, K. 1973. 'Gleichgewichtsverhältnisse bei der Messingherstellung nach dem Galmeiverfahren', *Erzmetall* 26:229–33.

Hawthorne, J. G., and C. S. Smith. 1979 [1963]. *Theophilus: On Divers Arts. The Foremost Medieval Treatise on Painting, Glassmaking and Metalwork.* New York: Dover.

Hoover, H. C., and H. L. Hoover. 1950 [1912]. *Georgius Agricola: De Re Metallica. Translated from the First Latin Edition of 1556.* New York: Dover.

Kibre, P. 1944. 'An alchemical tract attributed to Albertus Magnus', *Isis* 35:303–16.

Long, P. O. 1991. 'The openness of knowledge: An ideal and its context in 16th-century writings on mining and metallurgy', *Technology and Culture* 32:318–55.

———. 2001. *Openness, Secrecy, Authorship: Technical Arts and the Culture of Knowledge from Antiquity to the Renaissance.* Baltimore: Johns Hopkins University Press.

Martinón-Torres, M. 2001. *The Technology of Brass Production in Central Europe from the 10th to the 16th Century: Archaeometry and History.* Dissertation for MSc in Technology and Analysis of Archaeological Materials. London: Institute of Archaeology, University College London.

Martinón-Torres, M., and Th. Rehren. 2002. 'Agricola and Zwickau: Theory and practice of Renaissance brass production in SE Germany', *Historical Metallurgy* 36:95–111.

———. 2005. 'Alchemy, chemistry and metallurgy in Renaissance Europe: A wider context for fire assay remains', *Historical Metallurgy* 39:14–31.

Mitchiner, M. B., C. Mortimer, and A. M. Pollard. 1986. 'Nuremberg and its jetons, c. 1475 to 1888: Chemical compositions of the alloys', *Numismatic Chronicle* 147:114–55.

Newbury, B. D., M. R. Notis, and D. E. Newbury. 2005. 'Revisiting the zinc composition limit of cementation brass', *Historical Metallurgy* 39:75–81.

Packe, C. 1689. *The Works of the Highly Experienced and Famous Chymist, John Rudolph Glauber, Containing Great Variety of Choice Secrets in Medicine and Alchymy, in the Working of Metallick Mines, and the Separation of Metals . . .* London: Thomas Milbourn.

Picon, M., M. Le Nezet-Celestin, and A. Desbat. 1995. 'Un type particulier de grands récipients en terre réfractaire utilisés pour la fabrication du laiton par cémentation', in *Actes du Congrès de Rouen de la Société Francaise d'Étude de la Céramique Antique en Gaule.* Marseille: Société Francaise d'Étude de la Céramique Antique en Gaule, 207–15.

Pollard, A. M., and C. Heron. 1996. *Archaeological Chemistry.* Cambridge: The Royal Society of Chemistry.

Ponting, M. J. 2002. 'Roman military copper-alloy artefacts from Israel: Questions of organization and ethnicity', *Archaeometry* 44:555–71.

Ponting, M. J., and I. Segal. 1998. 'Inductively coupled plasma-atomic emission spectroscopy analyses of Roman military artefacts from the excavations at Masada, Israel', *Archaeometry* 40:109–22.

Priesner, C. 2000. 'Ein Compositum von Natur und Kunst': Zur Technologie der Messingfabrikation im 18. Jahrhundert', *Der Anschnitt* 52:130–41.

Rackham, H. 1952. *Pliny: Natural History.* London: Heinemann.

Rehren, Th. 1996. 'A Roman zinc tablet from Bern, Switzerland: Reconstruction of the manufacture', in S. Demirci, A. M. Özer, and G. D. Summers (eds.), *Archaeometry 1994.* Ankara: Tübitak, 35–45.

———. 1999a. 'Small size, large scale: Roman brass production in Germania Inferior', *Journal of Archaeological Science* 26:1083–87.

———. 1999b. 'The same . . . but different: A juxtaposition of Roman and Medieval brass making in Central Europe', in S. M. M. Young, A. M. Pollard, P. Budd, and R. A. Ixer (eds.), *Metals in Antiquity.* BAR International Series, 792. Oxford: Archaeopress, 252–57.

———. 2003. 'Crucibles as reaction vessels in ancient metallurgy', in P. T. Craddock and J. Lang (eds.), *Mining and Metal Production through the Ages.* London: The British Museum Press, 207–15.

Rehren, Th., E. Lietz, A. Hauptmann, and K. H. Deutmann. 1993. 'Schlacken und Tiegel aus dem Adlerturm in Dortmund: Zeugen einer mittelalterlichen Messingproduktion', in H. Steuer and U. Zimmermann (eds.), *Montanarchäologie in Europa: Berichte zum Internationalen Kolloquium 'Frühe Erzgewinnung und Verhüttung in Europa' in Freiburg im Breisgau vom 4. bis 7. Oktober 1990.* Sigmaringen: Jan Thorbecke Verlag, 303–14.

Riederer, J. 1988. 'Metallanalysen Augsburger Bronze- und Messingskulpturen des 16. Jahrhunderts', *Berliner Beiträge zur Archäometrie* 10:85–95.

Sisco, A. G., and C. S. Smith. 1951. *Lazarus Ercker's Treatise on Ores and Assaying, Translated from the German Edition of 1580.* Chicago: The University of Chicago Press.

Smith, C. S., and M. T. Gnudi. 1990 [1959]. *The Pirotechnia of Vannoccio Biringuccio: The Classic Sixteenth-Century Treatise on Metals and Metallurgy.* New York: Dover.

Thornton, C. P., and C. B. Ehlers. 2003. 'Early brass in the ancient Near East', *iams* 23:3–8.

Ullwer, H. 2001. 'Messingherstellung nach dem alten Galmeiverfahren', *Erzmetall* 54:319–26.

Weeks, L. 2004. 'An analysis of Late Pre-Islamic copper-base artefacts from Ed Dur, U.A.E.', *Arabian Archaeology and Epigraphy* 15:240–52.

Welter, J.-M. 2003. 'The zinc content of brass: A chronological indicator?', *Techne: La science aus service de l'histoire de l'art et des civilisations* 18:27–36.

Werner, O. 1970. 'Über das Vorkommen von Zink und Messing im Altertum und im Mittelalter', *Erzmetall* 23:259–69.

Wyckoff, D. 1967. *Albertus Magnus: Book of Minerals.* Oxford: Clarendon.

Zacharias, S. 1989. 'Brass making in medieval western Europe', in M. L. Wayman (ed.), *All That Glitters: Readings in Historical Metallurgy.* Montreal: The Metallurgical Society of the Canadian Institute of Mining and Metallurgy, 35–40.

Zwicker, U., H. Greiner, K.-H. Hofmann, and M. Reithinger. 1985. 'Smelting, refining and alloying of copper and copper alloys in crucible furnaces during prehistoric up to Roman times', in P. T. Craddock and M. J. Hughes (eds.), *Furnaces and Smelting Technology in Antiquity.* British Museum Occasional Paper 48. London: The British Museum, 103–15.

CHAPTER 10

Archives and Urban Archaeology: The Fairbank Surveyors' Papers and Work on Brownfield Sites in Sheffield

ANNA BADCOCK AND DAVID CROSSLEY

ABSTRACT *Urban archaeology in Sheffield benefits from the archive left by the Fairbank surveyors working in the town between 1750 and 1850. It includes field books containing site surveys and quantity-surveyors' books giving details and costings of structures. There is also correspondence and maps in varying stages of completion. Redevelopment schemes have required PPG16 archaeological assessments and evaluations on sites surveyed by Fairbank. In the Riverside area in central Sheffield, surveys have been matched with the excavated steel works and mill watercourses, and they give details relating to court cases; the Jessop steelworks excavation in the lower Don valley has produced timbers relating to an earlier forge; the Little London site in the Sheaf valley has demonstrated that site names can be attached to more than one place in the landscape, emphasising the need for caution when interpreting documentary evidence. These cases demonstrate how surveyors' papers complement field archaeology in towns, clarifying excavated features that can in turn add precision to records sketched by surveyors in the field. The Sheffield archive has been found to be unique in the extent of its survival, although more fragmentary collections exist elsewhere, notably for Tyneside and Glasgow.*

The Fairbanks and Their Records

Eighteenth-century landowners and property developers relied on the skills of the surveyor to define the boundaries of their land and to set out new building grounds within them. This expertise had been developed over the previous 200 years, based on the needs of a market in property that had originally been spurred by the release of monastic and crown lands, by the growth of commercial agriculture, and its counterpart in the development of market towns (Harvey 1993). A new precision in survey was apparent by the end of the 16th century, assisted by skills related to navigation and chartmaking; the London plans drawn by Ralph Treswell demonstrate what could be achieved (Schofield 1987). By the 18th century, numerous provincial surveyors were working in the planning and execution of urban and agricultural development (Thompson 1968: 19–63). The firm of Fairbank of Sheffield was one such, in practice for 100 years from the second quarter of the 18th century, and distinguished from the rest by the level of survival of its records.

In the middle and later decades of the 18th century, the prosperity of Sheffield was based on its metal industries, notably the manufacture of edge tools and cutlery. These trades prospered with the development of markets at home, particularly in agriculture, and abroad, notably around the Atlantic and in the colonies. Sheffield's development was matched in its region: agriculture responded to the needs of a growing town, and the tradition of metal working in the area led to farmer-craftsmen building new workshops and enclosing their lands, and to the construction of numerous mills for the forging and grinding of tools on the upper reaches of the five rivers that meet in or near the town.

The Fairbank papers are a key source for understanding these developments and the archaeological evidence for them. They survive in quantity from the years 1753–1848 and record the work of four generations of surveyors. The first William Fairbank (ca 1699–1759) was a schoolmaster in Sheffield. His extant work is sparse before 1753, the year of the first surviving book in the Field Book (FB) series. There are earlier maps in other collections in Sheffield Archives (the location of all subsequently cited documents unless otherwise stated); the earliest is of Ughill (1730), a hamlet north west of Sheffield (Bra 138); a plan of Silver Street, Sheffield dates from 1745 (SheS 1078); Wortley Top Forge was surveyed in 1746, and the town of Penistone was measured at about this time. This plan is reproduced by Addy (1965), but the whereabouts of the original are not now known. The source of the first William's surveying skills is uncertain, perhaps developed as an adjunct to his teaching of trigonometry in his school. He may have learned field skills from others who are known to have worked in the locality at this time. Ralph Gosling made the earliest surviving map of Sheffield in 1736, contemporary and usefully comparable with Thomas Oughtibridge's View of Sheffield of 1737. Gosling worked for the Duke of Norfolk in 1738 (ACM S164). The Norfolk archives name four surveyors of the Duke's woodlands (ACM S160, 282): John

Hall (1710), Thomas Smith (1710), John Stanley (1711 and 1715/6), and Robert Wilson (1715/16).

The beginning of the main Fairbank archive in 1753 appears to coincide with the entry of William Fairbank's son, also William, into the practice. The earliest survivor in the series of Field Books, the core of the collection, dates from this year, and henceforth we find numerous maps, in various states of completion, and other series of survey books. The Field Books contain surveys from whole estates to individual farms, houses urban and rural, roads, public water systems, and industrial buildings. There are occasional sketches of building elevations, and vignettes of buildings in their landscapes (Crossley, Richardson, and Unwin 2007). These books, and a related series – the Miscellaneous Books – form an unrivalled source for the topography of the region, containing names and holders of fields and plots and identifying their uses and those who were involved in con-struction. They form an essential and coherent source for the building of domestic and industrial property in Sheffield, for the development of the turnpike road system and for the improvement of estates and the enclosure of farms. The Early Field Books contain quantity surveys, but such was the extent of this side of the business that William Fairbank II started a separate series, the Building Books, in 1765.

Correspondence relating to the Fairbanks' 18th-century work does not survive. However, for the 19th century there is a copious collection for the decades when the third and fourth generation continued the business. Two of the 12 children of William II were involved, William III (1771–1846) and Josiah (1777–1844), joined by one member of the fourth generation, known in the papers as William Fairbank Fairbank (1804–1848). Changes in the archive in the 19th century reflect those in the local and wider economy, and in the technology of industry and transport. Work on road schemes was supplemented by canal contracts, and after 1820 there was a significant amount of railway business. After the pause in urban building at the time of the Napoleonic wars, Sheffield resumed its expansion, and there is important material on the layout of new streets in the 1820s and work connected with the buildings that lined them. The character of industrial building reflected the gradual change from water to steam power at this period. For example, in deciding on the form of new factory developments, clients needed information about the power produced by water wheels and steam engines. Disputes between neighbouring water-mill owners required particularly precise assessments by surveyors called in as arbitrators. Railway work, particularly on the new main lines of the 1830s, needed assessments of locomotive power required on gradients. There is evidence from the correspondence, from the 1820s onward, that the Fairbank office found difficulty in meeting these needs. There are complaints by clients about slow work, and there appears to have been confusion between the partners about procedure and accounting. The deaths of three partners in the 1840s ended the history of the Fairbank firm. Even without this turn of events, a lack of interest by the fourth generation and competition from local surveyors such as Marcus Smith might well have led to the end of the business.

The survival of the archive is in large part due to Marcus Smith, who purchased a major part of the office records, other than those sold to past clients by William Fairbank's widow. Smith would at first find the stock of maps useful, reusing surveys in much the same way that the Fairbanks had copied their earlier work. However, by the time Smith's nephew, the surveyor Alfred Denton, took over the stock, the topography of the town had changed sufficiently to reduce much of the collection to antiquarian rather than practical interest. Subsequently, the historical significance of the Fairbank papers was recognised, and they became part of the City Library collection in 1933.

Although the Sheffield archive material has been found to be unique in the extent of its survival, more fragmentary examples exist elsewhere, notably around Tyneside and in Glasgow, as well as, more generally, finished maps supplied by surveyors to their clients.

The Fairbank Papers and the Archaeology of Sheffield

In the 1980s, research on the water mills of the Sheffield rivers (Crossley et al. 1989) used the Fairbank papers as a key source. It was found that the existing indexes, some dating back to the time of public deposit, were insufficiently comprehensive, particularly in searches for individual plots and landholders and for persons working in the construction of property. A computer indexing project was undertaken, with the Field Books, containing around 20,000 items of survey, as the core. Cross references to other parts of the collection, the Miscellaneous Books, Accounts, Correspondence, or Apprentices Work Diaries were added as resources and volunteer-numbers permitted, and is ongoing. An interim Field Book index, in Access format, is deposited in the Sheffield Archive Office, although the development of a simple user interface remains to be done. What has become clear during the Fairbank project is the potential offered for the identification of features and buildings recorded during archaeological work.

The Fairbank material can be related to other 19th-century sources, notably the street directories and poor-rate assessment books, and forward comparisons can be made with the Goad fire-insurance plans. The company produced fire-insurance plans from about 1885 until 1970. The plans give details of land use, internal and external building construction, height, street width, building names, and addresses. Domestic dwelling and industrial activity/processes are also identified on the plans, giving excellent snapshots of the character of the urban environment.

This archive work has coincided with a remarkable increase in developer-funded archaeological work in Sheffield, brought about by the 1990 planning guidance note (PPG16), which ensures that archaeological matters are now a material consideration in the planning process. The city is typical; archaeological investigations nationwide have increased in number substantially as a result of this requirement.

Sheffield is undergoing something of an 'urban renaissance'. There is currently massive investment in the infrastructure of the city for both the public and private sectors. New offices, housing, retail quarters, and roads are being built or planned. Much of this development is inner city housing, and national government targets require that, by 2008, 60% of new housing should be built on brownfield sites (DETR 2000). Urban regeneration agencies are also highly active, bringing more brownfield sites into redevelopment schemes. This has led not only to a large increase in the number of archaeological investigations requiring both archive research and excavation but also to an increase in archaeological work on sites that have previously been developed and that are therefore highly likely to have a degree of archaeological and cultural historical interest. For the first time in Sheffield, archaeologists have begun to investigate the historical development of a city that was undergoing dramatic changes. Much of the new development has focused on river valley brownfield sites on the periphery of the city centre, precisely the locations that were developed in the 18th and 19th centuries to harness water power, and that the Fairbanks subsequently mapped.

Among the best examples are the Riverside area in the centre of Sheffield, where surveys have been correlated with the excavated steel works and the mill watercourses; the projected northern relief road across an area that was surveyed on various occasions in the 18th century; and archaeological investigations at the former Little London Works. These and other cases demonstrate the importance of surveyors' papers in guiding and complementing field archaeology in towns.

A lack of previous archaeological work in the city means that we must rely heavily on archive resources as part of our initial data-gathering exercises. The Fairbank material is just one of a suite of sources available, but it is invaluable for various reasons.

First, the collection is useful on a variety of different scales, from broad, citywide overviews to details about individual buildings and plots. Because they were intimately bound up with the development of the city and its infrastructure at that time, many of the smaller-scale surveys can be placed in a wider context of growth and investment. We are able to characterise the cityscape and understand how, and where, the city was spreading. The Fairbank city map of 1771 (Figure 10.1) shows proposed street layouts that are seen, realised, on early 19th-century maps and on the first OS map of the city of 1853.

Second, characterisation of the archaeological resource is also possible at a small scale, and some of the individual building and land holding surveys provide accurate and detailed information that helps us to target our archaeological investigations. This can be crucial when sites (notably steelworks) underwent subsequent major transformations with the development of late 19th- and early 20th-century technology. Archaeological investigations have demonstrated that large amounts of dumped material and industrial debris were frequently brought in, or moved around, in order to raise ground levels for modern constructions (for example, Badcock and May 2003; Symonds and Lines 2000). Archaeological remains may therefore lie buried beneath significant depths of 'made ground', and any map evidence that exists can help to refine the evaluation techniques and help establish a research framework around which to base investigations. The three short case studies given below give an example of the part played by the Fairbank archive in archaeological investigations.

Riverside Exchange

The Riverside Exchange development lies on the southwest bank of the Don, immediately upstream from the confluence with the river Sheaf, in whose angle lay Sheffield castle and the core of the medieval town (Figures 10.1 and 10.2). A lengthy programme of archaeological investigations has been undertaken in advance of, and during, the development of the site for offices and housing. The site is a fine example of historic riverside development in Sheffield; documentary evidence suggests that a town corn mill was established there in the 12th century (Crossley et al. 1989), surviving until the 19th century, when it developed into a brewery, whose successor was demolished for the current development. In the post-medieval period, there were grinding workshops (known in Sheffield as 'wheels'), a tannery and, from late in the 18th century, a steel works (Mulville et al. 1996). A number of Fairbank surveys exist for this area and have been significant in the interpretation of archaeological remains. These surveys date from the 1770s to the 1840s, and the associated field notebooks are present for the later maps of the series (Figures 10.3 and 10.4).

Figure 10.1 • Extract from Fairbank map, 1771; proposed street layouts can be seen at Alsop Fields (bottom centre); Millsands (detailed in Figure 10.3) can be seen adjacent to the cartouche at the top of the map (Sheffield Archives Fairbank Collection SheS 1S; reproduced by kind permission of the Head of Leisure Services, Sheffield City Council).

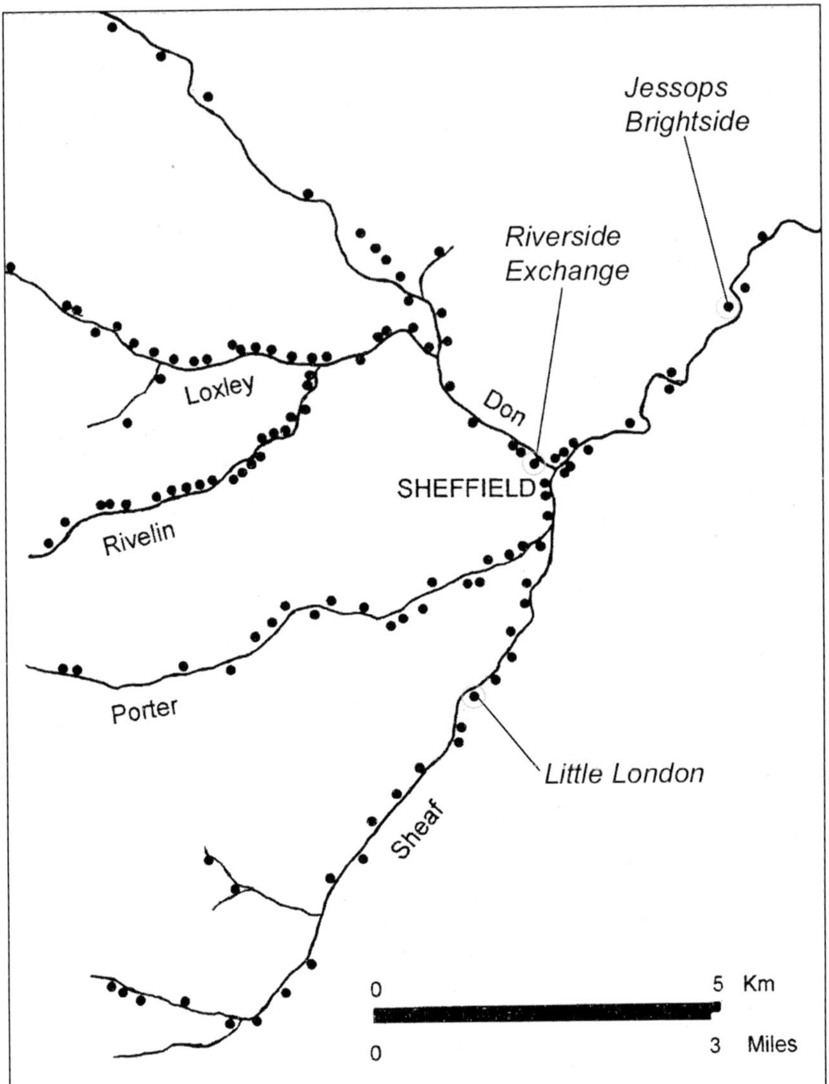

Figure 10.2 • Location of the case study sites, in relation to other early water-powered sites on the Sheffield rivers.

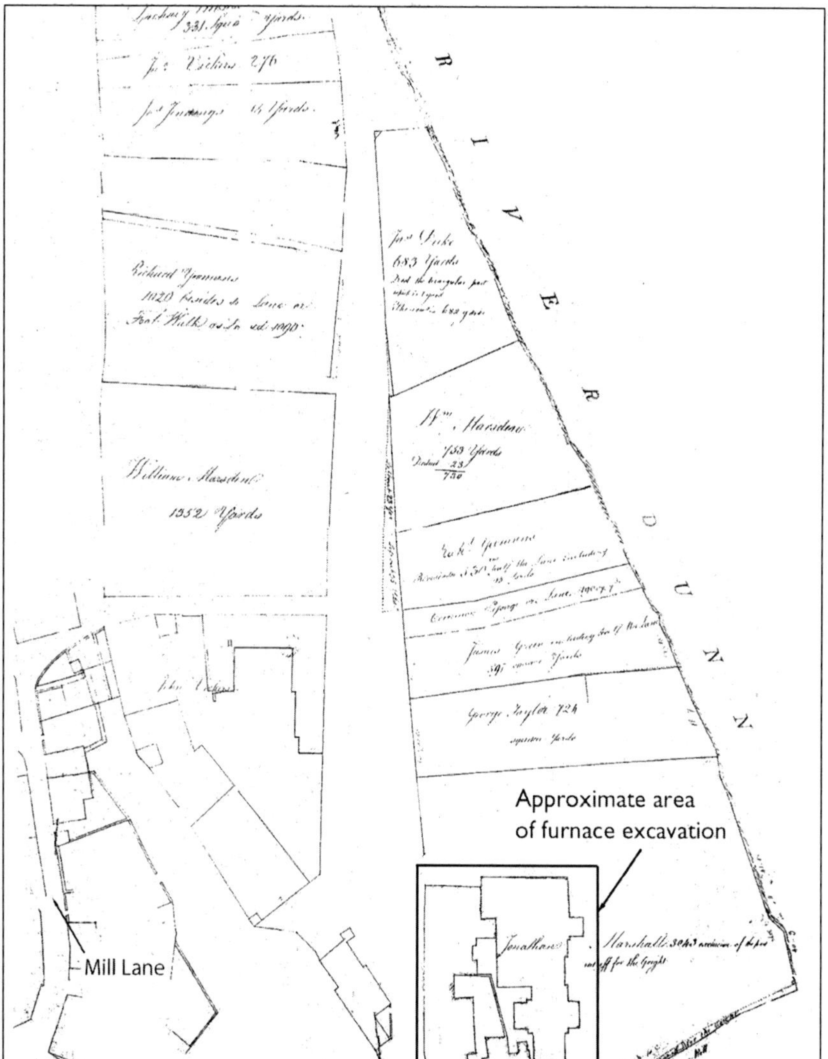

Figure 10.3 • Extract from Fairbank map, 1787, with our annotations (Sheffield Archives Fairbank Collection SheS 717L; reproduced by kind permission of the Head of Leisure Services, Sheffield City Council).

Figure 10.4 • Excavating the cutlers' wheel pit at Riverside Exchange (© ARCUS).

The 1787 Fairbank map (Figure 10.3) shows that the site was mainly open ground, except for the area around Mill Lane and Marshall's steelworks. The Marshall site plays a key role in the history of the steel industry in Sheffield. John Marshall had been in the steel trade since the mid-18th century, being steelmaker for the Cutlers' Company in 1759–1760. Soon after, he acquired his own site by the river and operated a successful crucible steel business. John Marshall died in 1793 and was followed by his nephew Jonathan, who continued until 1829. During this period Marshall's was the only establishment in Sheffield consistently de-scribed as 'converters and refiners' – implying the use of both cementation and crucible furnaces by this time (Mulville et al. 1996). A single-chest cementation furnace was described at this site by an industrial commentator in 1796 (Belford 1996), and archaeological excavations revealed that the foundations of several furnaces were pre-served in excellent condition. The furnaces appear to have been early examples of a unique form in Sheffield. Subsequent excavations at other sites have demonstrated the general adoption of a standardised double chest furnace in the 19th century, when the steel industry saw a massive and rapid expansion. Marshall was something of a pioneer in the city, and the Fairbank maps of the late 18th century captured this area during a critical period in its growth and development; by the time the first-edition OS map was made in 1850, the entire riverside had been developed, and the Marshall works had been incorporated into the works of another company.

Excavations further north on the Riverside site revealed evidence for a complex series of small-scale workshops. Water, of course, was a particularly valuable resource, since it powered the forges, grindstones, polishing wheels, and many other pieces of equipment. A Fairbank map of approximately 1828 is useful here, as it provides good detail about various buildings on the site, including the location of the wheel pits and a dam (mill pond). From this, we can establish who had access to water power and how it might have been shared among different occupiers. Water rights were critical to many businesses, and the capacity of the river could affect the design and layout of the wheels, and the pattern of the working day. The maps also document the increasing value of land, linked to the 'removal' of water from the townscape; many previously open goits and channels were turned into enclosed culverts that could be built on.

The archaeological excavations showed the wheel pits to have accumulated a wide variety of objects that were being produced in a number of different workshops. Cutlery and edge-tool discards, wire, pins, and rivets were among the items recovered from the bottom of the wheel pits (Figure 10.4). Many of these items showed evidence for being unfinished and appear to be the result of a long-held Sheffield habit of discarding damaged or faulty items in the river or water channels adjacent to the workshop (K. Hawley pers. comm.). The water moved much more slowly in the wheel pits than in the river itself and thus had a tendency to deposit items that had been swept into the wheel race rather than wash them downstream.

The map of ca 1828 is of particular interest. It was produced by the Fairbank office as evidence in a court case between Vickers (who were steel makers) and Gallimore who worked an adjacent grinding shop. The case involved a boundary dispute regarding the erection of a new building, with the plaintiff complaining that this blocked light to their workshop. A wall of the new building was torn down in the dispute, and this is recorded carefully on the Fairbank plan (Figure 10.5). Further detail is provided in the fieldbook, which records elevation details of the offending building (Figure 10.6).

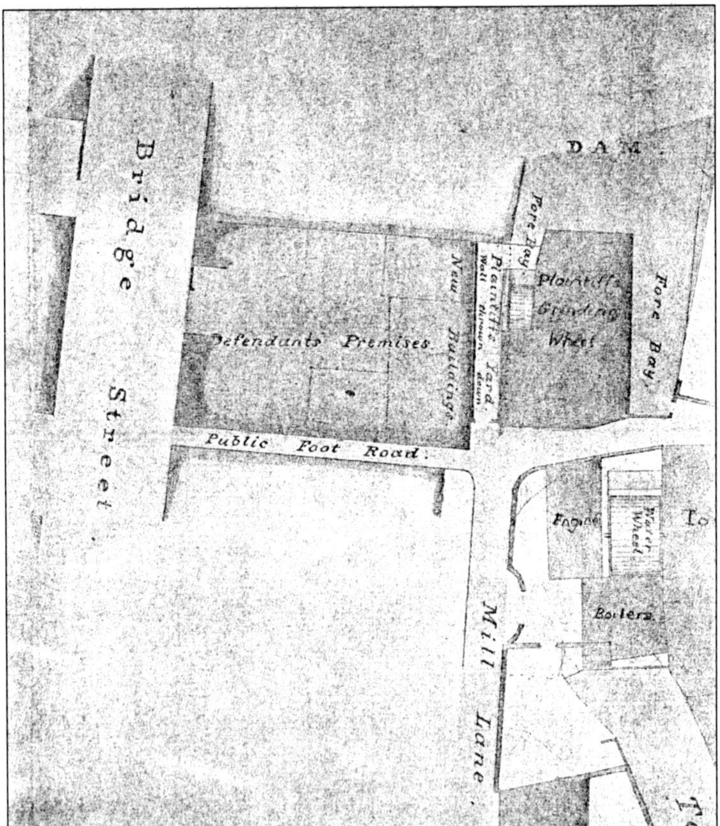

Figure 10.5 • Extract from Fairbank map 1828; note 'wall thrown down' in the Plaintiff's Yard, adjacent to 'Plaintiffs Grinding Wheel'. The 'New Buildings' in the Defendant's Premises are those shown on Figure 10.6 (Sheffield Archives Fairbank Collection FB SheS 7145; reproduced by kind permission of the Head of Leisure Services, Sheffield City Council).

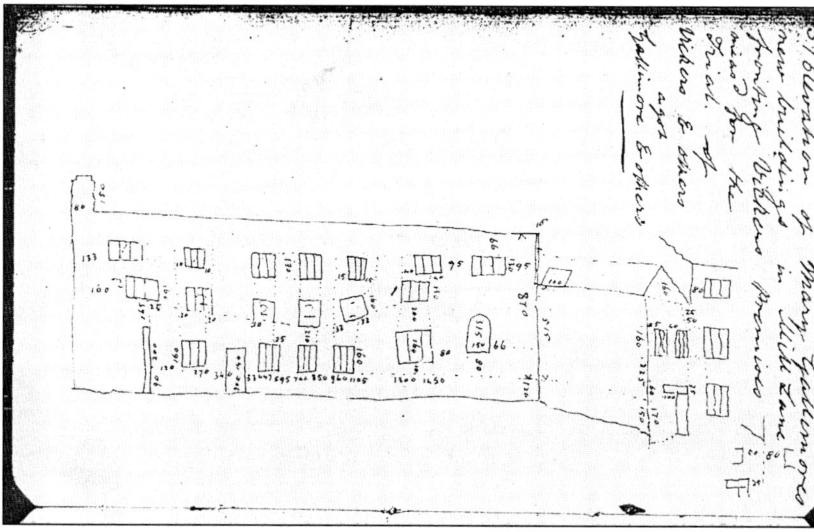

Figure 10.6 • Page from a Fairbank Fieldbook, showing detail of elevation, relating to dispute (see Figure 10.5) (Sheffield Archives Fairbank Collection FB191; reproduced by kind permission of the Head of Leisure Services, Sheffield City Council).

Altogether, this source not only gives us invaluable detail about property, landholding, and the functions of certain buildings but also gives us insights into relations between the different families and businesses working here, which could not be gained from the archaeological record.

Jessop's Brightside Works

Like the Riverside Exchange, the former Brightside Works was located on the banks of the River Don, but further down river, to the northeast of the city centre. Here, too, a medieval corn mill site was developed into a cutler's wheel late in the 16th century. Several other wheels had been added by the early 18th century, and by 1738 the first reference to a 'tilt' in Norfolk rental papers suggest a forge was operating on the site (Crossley et al. 1989). During the 18th and 19th centuries, the forge expanded rapidly to become a major steel works, with an associated slitting mill on the opposite side of the Don. The weir provided a head of water for both sides of the works.

The Fairbank map of 1789 (Figure 10.7) shows the forge, associated buildings, dam (mill pond), and weir before the expansion that took place prior to the OS mapping of 1854.

The location of the dam proved to be very significant archaeologically, because although no structural remains of the original forge or corn mill

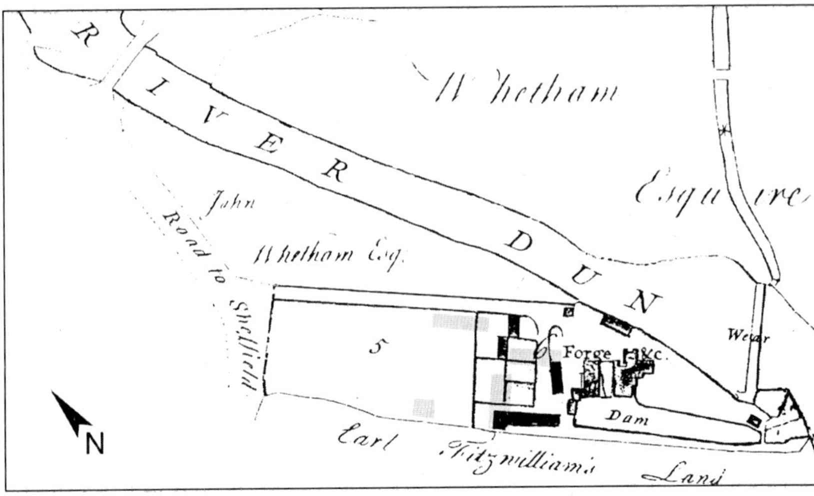

Figure 10.7 • Extract from Fairbank map, 1789; location of trenches indicated by grey shading (Sheffield Archives Fairbank Collection FB SheD 788; reproduced by kind permission of the Head of Leisure Services, Sheffield City Council).

buildings were located *in situ*, over 100 massive timbers were preserved in waterlogged conditions in the (infilled) dam (Figure 10.8). These were identified as the remains of tilt hammers, anvils, and structural elements of the forge itself. The majority were oak and were therefore suitable for dendrochronological dating, and a range of felling dates between approximately 1714 and 1850 were obtained from 16 samples (Tyers 2001).

Photographic evidence from the early 20th century (Figure 10.9) illustrates tilt hammers still in use (under water power), showing that some elements of the older forge buildings on the Fairbank plan were incorporated into the steelworks in the mid-19th century. Exactly when the rest of the older buildings, remaining tilt hammers, and framework were deposited in the dam is uncertain, but between 1935 and 1954 the dam was filled in, allowing the timbers to be preserved in the wet silts that had accumulated at the bottom.

Similar examples of continuing use of technologies were encountered in other parts of the site. One of the excavation trenches revealed the remains of a workshop that was appended to a building shown on the Fairbank plan. Internal wall divisions provided good phasing evidence and showed that although the space had been expanded, it continued to remain divided into small working zones with individual hearths, chimney bases, and possible casting areas, all of which would have provided space for only one or two people at a time. Thus, the archaeological evidence

Figure 10.8 • Selection of timbers recovered from the infilled dam; identifiable elements include anvils, anvil bases, hammer arms, and structural timbers (metre scale toward centre) (© ARCUS).

Figure 10.9 • Tilt hammers still in use ca 1911; reproduced from *A visit to a Sheffield Steelworks*. These last surviving hammers have been preserved at the Abbeydale Industrial Hamlet, in Sheffield.

demonstrates the endurance of small-scale production and finishing activities at an individual level, despite the ever-increasing volume of production and physical expansion of the works.

Little London Works

The former Little London Works, operated by W Tyzack, Sons, & Turner, was a mid-19th century steelworks, producing mainly scythes, saws, and agricultural machine parts, built on the site of a late 18th-century tilt forge and grinding shop, the Little London Wheel. The forge was located on the river Sheaf, approximately 2.5 km south of the city centre.

The origins of the forge are unclear; lease papers show that the Little London Wheel was established by 1720, and although there are no pre-18th century sources of evidence for this wheel, it is possible that it was previously known as New Mill, for which 16th- and 17th-century references exist, but no precise location is given (Crossley et al. 1989:102).

If the wheel did have pre-18th-century origins, the map evidence suggests that it was in a different location. The earliest map of this stretch of the river is a Fairbank plan of the nearby Heeley Mill in 1770 (Figure 10.11), and although an area is marked 'Little London' (top left), this is not the site of the 18th-century forge, which at this time appears to

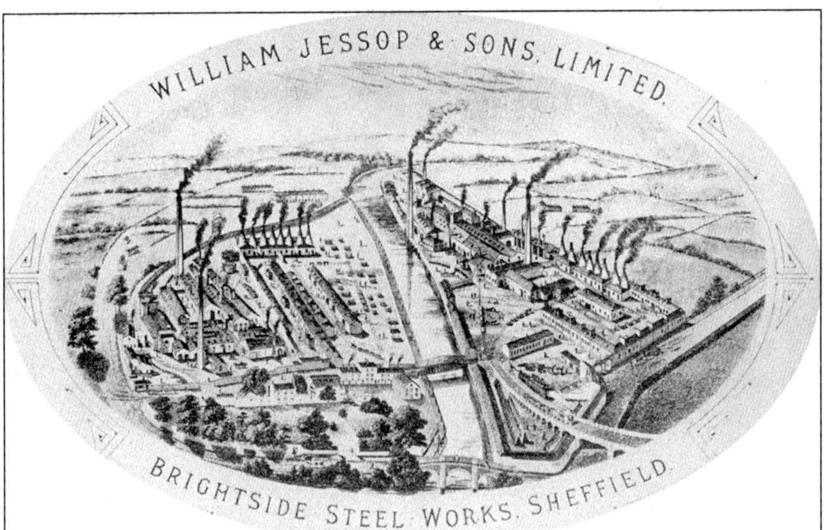

Figure 10.10 • Etching of the Brightside Works ca 1879; the excavations investigated several of the buildings shown on the right hand side of the image. Note the conical chimneys of two sets of cementations furnaces (reproduced from *The Sheffield Illustrated Guide*, 1979).

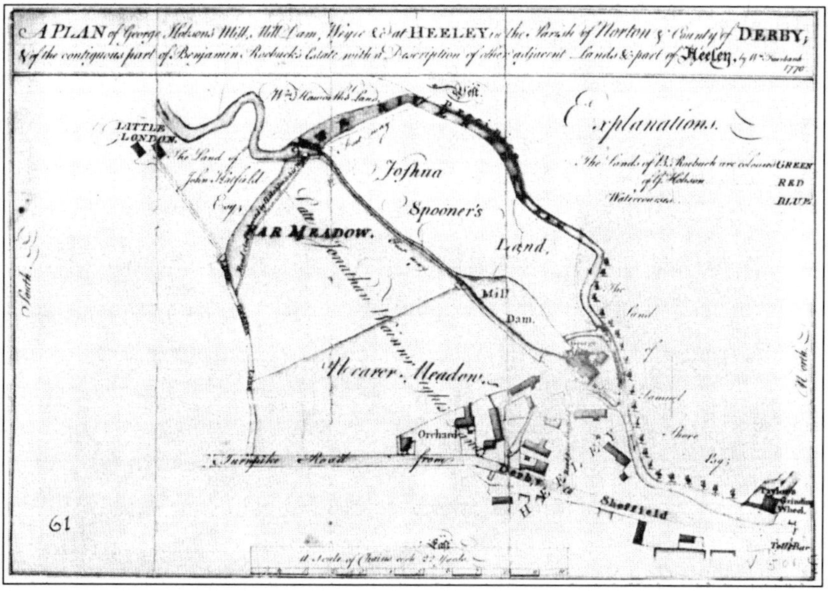

Figure 10.11 • Fairbank map of 1770; The Little London Wheel was later built adjacent to the river, above the words 'Far Meadow' (top left) (Sheffield Archives Hobson Collection HC 61; reproduced by kind permission of the Head of Leisure Services, Sheffield City Council).

be undeveloped. The forge was first mapped in 1787, with an associated dam. The new dam may have masked any trace of an earlier wheel slightly further upstream, and it would certainly have covered the two Little London buildings shown on the 1770 map. This highlights the difficulty in attempting to trace sites by their name alone through documentary sources, since sites may be renamed, or a name may have had a somewhat fluid attachment to place in the past.

The 1787 Fairbank map helped to determine which of the structures at the 19th-century wheel were the earliest, as by the time of the next map of the area in 1849, the works had already expanded. The quality of the Fairbank mapping allowed building footprints and watercourses to be plotted and targetted for excavation, the latter being particularly important, since we have demonstrated that deep channels and wheel pits are the prime areas for the accumulation of artefacts at these types of sites.

Archaeological excavations demonstrated that the dam wall was still intact below ground, and in parts survived above ground, incorporated into a later revetment wall. The pits of both water wheels that powered the tilt hammers in the northern building and the grinding shop in the southern building were uncovered. The tilt hammers were still present, although derelict, in the 1970s (Figure 10.12), but the entire steelworks

Figure 10.12 • The Little London tilt hammers in the 1970s, just before demolition; the hammers were donated to Beamish Open Air museum (© D. Crossley).

was demolished soon after, and the land built over, masking any sign of the previous structures prior to their excavation. The wheel pits contained debris from the demolition, but also artefacts that had been discarded during production when the forge and grinding shop were operational. Analysis of these artefacts is ongoing, but several stages of the production process are represented.

Conclusion

The Fairbank archive is a valuable source of information for anyone investigating the historical development of Sheffield. The value of the collection for archaeologists lies in the combination of field notebooks and finished maps, which help us to set sites within the wider context of a rapidly developing cityscape and to target our investigations on a site-by-site basis. The notebooks provide background information that was often omitted from the final 'finished' maps. The surveys have helped to shed light on the development of key sites in the growth of the Sheffield steel industry and can occasionally give glimpses into the working lives of people at the time. The archaeological evidence adds depth and detail to this information and encourages us to critically re-examine archive evidence.

Although maps, like any written source, are subjective, they are a useful tool for directing our investigations and for enhancing interpretations and understanding of our built and buried environments.

Acknowledgments

Acknowledgments are extended to Andy Lines, for providing information on excavations at Riverside Exchange, to James Symonds and Mark Edmonds for commenting on this text, to the staff of the Hawley Collection for sharing their vast knowledge of the edge tool and cutlery industries, the staff of the Sheffield Record Office and Local Studies Library, and especially to all those who helped to excavate the sites that are discussed.

References

Archive sources

Fairbank Map of Sheffield [Sheffield Archives SheS 1S]
Fairbank Map 1770 (for Little London) [Sheffield Archives HC 61]
Fairbank Map 1787 (for Riverside) [Sheffield Archives SheS 717L]
Fairbank Map 1787 (for Little London) [Sheffield Archives Nort 27L & FB63]
Fairbank Map 1789 (for Riverside) [Sheffield Archives SheD 788]
Fairbank 1828 Vickers v Gallimore (for Riverside) [Sheffield Archives FB SheS 7145]
Fairbank Fieldbook (for Riverside) [Sheffield Archives FB191]

Publications

Addy, J. (ed.). 1965. *A Further History of Penistone.* Penistone: Penistone WEA History Group.

Badcock, A., and R. May. 2003. *Archaeological Evaluation of Land Off Wellington Street and Eldon St, Sheffield.* ARCUS unpublished interim report 742b.

Ball, C., D. Crossley, and N. Flavell. 2007. *Water Power on the Sheffield Rivers* (2nd, rev. ed.). Sheffield: Sheffield Trades Historical Society.

Belford, P. 1996. *An Archaeological Evaluation at the Former Marshall's Steelworks Site, 'Riverside Exchange', Sheffield.* ARCUS unpublished report 240b.

Crossley, D., J. Cass, N. Flavell, and C. Turner. 1989. *Water Power on the Sheffield Rivers.* Sheffield: Sheffield Trades Historical Society.

Crossley, D., M. Richardson, and J. Unwin. 2007. 'Landscapes and buildings: A neglected aspect of the work of the Fairbanks of Sheffield', in M. Meek (ed.), *The Modern Traveller to our Past.* Newtownards: Colourpoint.

Department of Environment. 1990. *Planning Policy Guidance Note 16 (PPG16).* London: Archaeology and Planning.

DETR [Department for Environment, Transport and the Regions]. 2000. *Our Towns and Cities: The Future – Delivering an Urban Renaissance* (Urban White Paper), Department for Environment, Transport and the Regions, Cm 4911.

Harvey, P. D. A. 1993. *Maps in Tudor England.* London: The British Library and Public Records Office.

Mulville, J., G. Stroud, G. Davies, and J. Symonds. 1996. *Desk-Based Assessment of Riverside Exchange, Sheffield.* ARCUS unpublished report 240.

Pawson, H., and J. Braislford. N.d. *Visit to a Sheffield Steelworks.* Sheffield: Jessop W. & Sons Ltd.

———. 1862, *The Illustrated Guide to Sheffield.* Wakefield.

Schofield J. (ed.). 1987. *The London Surveys of Ralph Treswell.* London: London Topographical Society, 135.

Symonds, J., and A. Lines. (eds.). 2000. *Riverside Exchange – Interim evaluation report,* ARCUS unpublished report 240d.

Thompson, F. M. L. 1968. *Chartered Surveyors: The Growth of a Profession.* London: Routledge.

Tyers, I. 2001. *Dendrochronological evaluation of timber samples: Jessops Riverside,* ARCUS unpublished report, 595b.

About the Authors

Sally-Ann Ashton studied classical archaeology at King's College London and was awarded a PhD in 1999. She worked in the Department of Greek and Roman Antiquities at the British Museum as co-curator of the special exhibition *Cleopatra of Egypt: From History to Myth*, before working at the Petrie Museum of Egyptian Archaeology, University College London. She is currently Senior Assistant Keeper in the Department of Antiquities at the Fitzwilliam Museum, Cambridge. She has published extensively on many aspects of Ptolemaic and Romano-Egyptian material culture, including the following monographs: *Ptolemaic Royal Sculpture from Egypt: The Greek and Egyptian Styles and Their Interaction* (1999); *The Last Queens of Egypt* (2003); *Petrie's Ptolemaic and Roman Memphis* (2003); *Roman Egyptomania* (2004); and *Cleopatra*, with S. Walker (2003). She is Field Director of the North Karnak Survey, and her current research interests include settlement at North Karnak and the role of museums in prisons.

Anna Badcock is Assistant Director of Archaeological Research and Consultancy at the University of Sheffield (ARCUS). Anna graduated from The University of Sheffield in 1992 and since then has developed a profile in the archaeology of industry and brownfield regeneration. She is a member of Council for the Society for Post-Medieval Archaeology and sits on the Archaeology Committee of the Historical Metallurgy Society.

Justine Bayley was awarded a PhD from the University of London for research into nonferrous metalworking in England. She works for English Heritage, leading its Technology Team, publicising and popularising the benefits of using scientific techniques to aid the interpretation of archaeological finds. Her publications include books on Anglo-Scandinavian metalworking in York and on the technical study and analysis of Roman brooches. Her current research focuses on northern European metal- and glassworking of Roman and later date. She has jointly edited the journal *Historical Metallurgy* since 1990.

David Crossley is retired (now Honorary) Reader in Archaeology at Sheffield University. He has undertaken archival and archaeological research on postmedieval iron and glassmaking sites, notably blast furnaces and forges in Yorkshire, the Weald, and in Italy, and glass furnaces in Yorkshire and the Midlands. He has contributed to the English Heritage Monuments Protection Programme. His current research interests are in the archaeology of water power and the role of the surveyor in the British Industrial Revolution. He has jointly edited the journal *Historical Metallurgy* since 1990 and was editor of *Post-Medieval Archaeology* from 1979 to 1989. He chairs the English Heritage Industrial Archaeology Panel and is a Trustee of the Ironbridge Heritage Foundation.

Ian Freestone is Professorial Fellow and Head of Archaeology and Conservation in Cardiff School of History and Archaeology. He gained undergraduate and postgraduate degrees in geology before undertaking postdoctoral work in experimental petrology. He subsequently joined the British Museum as a scientist, where he undertook the analysis of artefacts from a wide range of places, periods, and cultures. He became Deputy Keeper of the Department of Conservation, Documentation, and Science in 2002 and joined Cardiff University in September 2004. He is currently investigating the early glass industry, using elemental and isotopic analysis. A specialist in early materials and technologies, he was awarded the Pomerance Medal of the Archaeological Institute of America for scientific contributions to archaeology in 2004. He is President of the Association for the History of Glass and on the editorial boards of *Archaeometry*, *Facta*, and *Journal of Archaeological Science.*

Carolyn Graves-Brown is curator of the Egypt Centre, Museum of Egyptian Antiquities, at University of Wales, Swansea. She holds a degree in archaeology from the University of Durham and the Museums Association diploma, and has worked in several museums since 1983. She is presently writing up her PhD on the ideology of Egyptian Dynastic lithics as a part-time student at the Institute of Archaeology, University College London, and has published several articles on Egyptian Dynastic flints.

Caroline Jackson is Senior Lecturer in Archaeological Sciences at the University of Sheffield. She specialises in experimental and analytical techniques in archaeology to interpret material residues and how they inform us about past societies. She has published most extensively in the field of archaeological glasses, studying material that dates from the Bronze Age to the postmedieval period, spanning a geographical area that covers North Africa through Europe to North America.

Marcos Martinón-Torres is Lecturer in Archaeological Science and Material Culture at the Institute of Archaeology, University College London. With degrees in history, archaeology, and archaeological science, he has a special interest in the integration of different research approaches to past materials and technologies and the way in which these inform about human theories, perceptions, and actions. Most of his research has focused on medieval and postmedieval metallurgy and crucible technology, with a strong emphasis on the study of alchemical and metallurgical laboratories, and the processing of noble metals. He is also interested in the transfer of material culture and ideas between indigenous peoples and Europeans in colonial contexts. He currently works in projects in Europe, America, and China.

Carlos Pardos works as a laboratory technician at the Department of Analytical Chemistry of the University of Zaragoza, Spain. He completed his MSc in Chemistry in 1992 and his MA in the History of Art in 2000. He routinely combines his expertise in chemistry and art for the study of cultural heritage.

Josefina Pérez-Arantegui is Lecturer in the Department of Analytical Chemistry of the University of Zaragoza, Spain, since 1993. She received her PhD in Analytical Chemistry from the University of Zaragoza in 1991 and conducted postdoctoral research at the Centre de Recherche et de Restauration des Musées de France (Louvre Museum, Paris, France). Her research interests focus on archaeological chemistry, especially in the analytical chemistry of ancient inorganic materials (ceramic, glass, glaze, and metal) and in the application of new analytical techniques. Her publications have appeared in various journals related to analytical chemistry in cultural heritage, archaeometry, and archaeological sciences. She has also authored several book chapters and articles in the *Encyclopedia of Analytical Science* and the *Encyclopedia of Archaeology.*

Thilo Rehren is Professor for Archaeological Materials and Technologies at the Institute of Archaeology, University College London. His academic background includes a first degree and PhD in mineralogy and petrology, and a habilitation in archaeometallurgy. He is particularly interested in reconstructing high-temperature production processes for metals and glass and in the interplay between natural, system-driven aspects of technology and the role of human choice and activity in mastering past technologies. He has worked extensively on a range of metals, specifically lead and silver smelting, silver refining, crucible technology in general, Islamic crucible steel production, Roman zinc and Roman and medieval brassmaking, and early platinum, tin, and tungsten production. More recently, he has jointly with Edgar Pusch identified the first firm evidence for Egyptian glassmaking, at Qantir-Pi-Ramesse in the Nile Delta. He is Editor of the *Journal of Archaeological Science* and currently serves as the President of the Society for Archaeological Sciences.

Andrew J. Shortland is Senior Lecturer in the Department of Materials and Applied Science at the University of Cranfield and Director of the Centre for Archaeological and Forensic Analysis. The Centre specialises in applying scientific techniques in novel ways to solve forensic and archaeological problems. Dr Shortland specialises in the analysis of glass, glaze, and ceramics and the application of trace element and isotopic techniques to the provenancing of objects and people.

Jim Smedley is Departmental Superintendent in the Department of Engineering Materials, Sheffield University. He has over 30 years experience of experimental laboratory scale glassmaking investigating the physics and chemistry of glasses and the glassmaking process. The application of these techniques is used to observe the glass-forming characteristics of raw materials and their mixtures thus allowing us to assess the capabilities of glassmakers of early societies. The preparation of raw materials also features strongly in this work. His main interest is in the glassmaking industry of medieval and postmedieval northwestern Europe, and this is reflected in his recent publications.

Index